Pollution, property & prices

Pollution property &prices

An essay in policy-making and economics

JHDales

UNIVERSITY OF TORONTO PRESS

Preface

This is a book about pollution *problems*, not about pollution itself. It contains virtually no factual information, and very little in the way of outraged denunciation of evil. I have attempted, instead, to identify those aspects of pollution problems that make them so perplexing and so difficult to control; and toward the end of the book I do come up with one suggestion that I hope offers a practicable way of coping with them.

The argument of the book is constructed from the viewpoint of economics, and the final product might appropriately have been entitled "An Economist Looks at Pollution." The branch of economics that is relevant to pollution problems is concerned with *social* problems and *social* decision-making; it is a relatively new branch of the subject and is less well developed than the analysis of *individual* problems and *individual* decision-making that is the main concern of traditional economics. My own foray into this new area in pursuit of pollution problems has led me to a vivid awareness of the very close relationship between law and economics. Specifically, the linkage is between prices – the stuff of economics – and the law of property, or more explicitly the law of property *rights*. Property rights constitute the set of social rules that on the one hand gives individuals the right to use their "property" in certain ways and on the other hand forbids them to use it in other ways. (It is very important to grasp the idea, expressed at the beginning of chapter v, that you never own *things* but only *rights* to the *use* of those things. You may own your house "outright," but you still have no right to burn it

v

down.) In recent years other writers (some of them mentioned in the note on further reading at the end of the book) have dealt with the theoretical aspects of the relationship between law and economics; the present book is, in one sense, an attempt to apply their findings to the pollution field. It is this line of thought that leads to my "economico-legal" proposal for dealing with pollution problems.

This work derives from a study of the economics of pollution undertaken during a sabbatical leave from my teaching duties at the University of Toronto, and has been financed in part by the University and in part by a Senior Fellowship from the Canada Council. Several of my colleagues have listened patiently to my evolving views on pollution problems and their comments have helped me sharpen those views. I must acknowledge especially the help I have received from John McManus, who has taught me much about the relationship between property rights and economics. I have benefitted greatly, too, from several discussions with Dr. A. D. Misener, Director of the Great Lakes Institute at the University of Toronto; like a growing number of physical scientists concerned with the study of pollution, he has a lively awareness of the fact that pollution problems are a complex amalgam of problems in physical science and problems in social science.

August 1968

Contents

To live is to pollute

THE QUANTITY OF WASTE

In the most general terms, the economic processes of a society may be thought of as a continuous flow of materials. Men take from the environment a wide variety of vegetable, mineral, and animal materials; transform them into a very much wider variety of economic goods; consume these goods, a process during which they undergo physical or chemical transformation and become, in effect, garbage; and then discard these unwanted products of consumption into the environment. Unwanted materials – solid, liquid, or gaseous – are also generated during the second phase of the cycle (the production process) and they too are discarded into the environment. In a society where nothing changes from one year to another – where population is constant, production and consumption in one year are exactly the same as in the previous year, and there is no technological change – the Law of the Conservation of Mass tells us that the annual tonnage of discarded, or waste, materials is exactly equal to the tonnage of natural materials – vegetable, mineral, and animal – taken from the environment. Although the tonnage of waste materials equals the tonnage of materials taken from the environment, the types and forms of the discarded stuff are very different from the natural materials that enter the system. We shall have more to say about the types of wastes later in this chapter, but for the moment we

1

shall pursue further the question of the *quantity* of wastes, measured in tons. (For the sake of simplicity, we shall conduct the argument of this book purely in terms of waste *materials*. It should be noted, though, that waste *energy*, especially in the forms of heat and noise, can also cause pollution problems; for example, enormous amounts of waste heat from electrical generating plants, both steam and atomic, are discharged into natural water systems, and sometimes lead to undesirable changes in aquatic flora and fauna.)

The principle we have just established about the total tonnage of wastes generated in a simple "stationary" or "circular flow" economy applies with only very slight modifications to actual, complex economies. As the result of exports and imports of goods (and therefore of wastes) actual economies may have to dispose of a slightly different number of tons of waste than the tons of natural materials they process. Moreover, in a society that is growing, in the sense that it uses more and more natural materials from the environment as time goes by, the inventory of materials in the form of goods or human bodies will be increasing, and therefore the annual tonnage of wastes, though increasing every year, will be slightly *less* than the annual tonnage of natural materials taken from the environment. There will also be other short-term variations in materials in inventory resulting from such things as changes in investment and obsolescence. But in the end, as is obvious, what is taken from the earth will, by one route or another, return to the earth – at least until we start shooting our garbage into outer space!

Let us now consider the factors that determine what the actual annual tonnage of wastes generated by a society will be. So long as the techniques of production and consumption – what economists refer to as a society's technology – remain unchanged, it is obvious that the total annual tonnage of natural materials used (or waste materials generated) will depend on the number of people in the economy and the per capita consumption of materials (or, roughly speaking, the standard of living of the population). In brief, the tonnage of wastes generated by a society varies directly with its population and its affluence.

2

Another important general principle about the amount of waste generated by a society is that it varies *inversely* with the level of the "technology of recycling" achieved by the society. Let us develop this principle carefully. In what I have said so far I have been careful to define waste as unwanted material discarded into the environment. If we think of the behaviour of individual firms and households, it is clear that not all of the stuff they want to get rid of is discarded into the environment. Materials that are surplus or unwanted by any one firm or household may be sold, or given away, if some other firm or household wants them. Waste, or material discarded into the environment, is made up only of those materials that no one wants, i.e., that no one will either pay for or accept as a gift. As man's skill in utilizing materials improves, materials that were formerly wastes may become valuable and therefore no longer discarded: sawdust becomes some form of composition board; old iron machinery becomes new steel ingots; unwanted smelter gases become fertilizer; empty bottles are re-filled; cinders are made into building materials; and so on. Thus the more thoroughly, and the more often, we can use and re-use materials, the less the waste that results from a given production (and consumption) of goods.

Technological changes that are materials-saving work in the same direction. Engines that are lighter per horsepower developed, electric plants that use less coal per kilowatt hour generated, and furniture that is sturdy though light, all illustrate materials-saving developments that result in less garbage per unit of consumer satisfaction. Thus while growth in population and affluence increases waste, technological improvement that either reduces the use of materials or reduces the use of new materials by recycling used materials, reduces waste. To put the same point in another way, better materials-saving and recycling techniques allow a society to increase its production (and consumption) with a less than proportionate increase in its demand on the environment for natural materials, and therefore a less than proportionate increase in the waste that it discards into the environment. The junk man, the antique dealer, and the scientist who finds new ways to reduce the annual tonnage of natural materials used in production, all

3

make important contributions to man's struggle to increase his consumption without being smothered by his wastes.

KINDS OF WASTES

Every material known to man is represented in his dumps, effluents, and emissions. Discarded materials may be classified in a great many ways, two of which are especially useful for our purposes. We first consider a classification of wastes according to their durability, or the length of time during which a waste material retains its original form and properties after it has been discarded into the soil, the water, or the air. We shall then say something about the amount of "damage" associated with different wastes (and different waste disposal practices).

Wastes differ dramatically in their durability. Human and animal wastes, vegetable matter, and so on, begin to be degraded by bacteriological action as soon as they are put into soil or water, and they normally disappear within a matter of weeks or months; typical products of these natural processes of degradation are stable inorganic materials such as nitrates, phosphates, and sulphates that serve as fertilizers for plant life. It is as a result of such action that water is often said to be "self-purifying"; water polluted at one point by sewage, for example, will cleanse itself and become "pure" again after a few days (and probably several miles of travel), the rate of recovery depending on several factors, especially the temperature and the aeration of the water. A similar process takes place in soil, as everyone knows.

What must be stressed, however, are the limitations of natural purification processes. Such processes do not prevent or avoid pollution. There is short-run pollution – a reduction in the value of the water or land for other uses – while the recovery process is taking place. And both soil and water will become continuously polluted whenever the rate of discharge of organic wastes exceeds the absorptive capacity of the particular piece of land or body of water concerned. Neither is there any "self-purification" with inorganic wastes. The main pollution problem that results from the use of detergents derives from the high phosphate content of these materials; the new "bio-degradable" detergents prevent the

4

formation of detergent foam in rivers and lakes, but they do nothing to reduce the "algae pollution" that results from the fertilizing effect of detergent phosphates. Nearly all air pollution, too, comes from inorganic wastes, so air is not self-purifying. Wastes discharged into the air are diluted by air currents, and may be changed into other forms through chemical processes; but they either remain indefinitely in the air or fall out of the atmosphere and contribute to soil or water pollution.

Other materials, some metals for example, are also "digested" by natural forces – rusting is an example – though only over periods measured in decades or even centuries. Still other materials, such as glass, and the tailings and slag heaps generated by mining and smelting operations, never change their form for all practical purposes. In the last few decades scientific advance has greatly increased the amount of "perpetual" waste that is dumped into the environment. Some radioactive wastes are so dangerous to life that they have to be kept out of circulation altogether, and some of these degrade so slowly that for practical purposes they have to be stored in perpetuity. Many of the new chemical wonders, such as plastics, and pesticides that are poisonous to many forms of plant and animal life, are very stable inorganic substances that break down only over prolonged periods of time; they therefore accumulate in soil, water, and living organisms (including you and me) as each year's production is added to the stock already in the environment.

From man's point of view, the harm done by discarding a waste into the environment often depends not so much on the properties of the waste itself as on other factors: the chemical and biological processes that take place *after* the waste has been discarded; and how and where it is discarded. We shall have much to say later on about the "how and where" of waste disposal, but one example here will illustrate the sort of point at issue. Animal manure, if spread over fields in appropriate quantities, will enrich the soil and be beneficial to crop production; if it is simply left to rot in unsightly and ill-smelling heaps, it may become a breeding place for disease-carrying insects and therefore a health hazard; if it is dumped into a stream or a lake, it may result in a serious pollution

5

problem and a reduction in the value of the water for recreational, domestic, and industrial purposes. In the first instance, the manure is not a waste at all. It is a fertilizer, an input into the farming production process, and its use illustrates a recycling technique in agriculture. If, however, the manure is not used by its owner and cannot be sold or given away to anyone else, it is a waste. Thus a given material may be a waste under some conditions (at some locations and some times) and not under others. Admittedly this may complicate the measurement of waste to some extent; but at least our definition holds: waste is something without value that no one "will either pay for or accept as a gift." An important corollary of this definition is that it will always cost something to "get rid of" a waste. How much it costs, as we shall see, depends on the "how and where" of its disposal.

Natural processes – chemical, physical, and biological – that occur *after* a waste is discarded are incredibly complex, and their variety is enormous. Probably only a small minority of them have been identified and analysed. Some waste substances interact to produce new substances, or new effects that differ dramatically from the effect of either substance alone. Neither the inhalation of fumes from carbon tetrachloride nor the presence of alcohol in the blood is particularly harmful in itself, but a person who drinks beer while he cleans a rug with carbon tetrachloride in a poorly ventilated room is likely to be dead within a few hours; the vapours from the cleaner interact with the alcohol in the blood to produce a deadly reaction. It is suspected that some of the chlorinated hydrocarbon pesticides are much more harmful to fish and mammals when there are two or more different kinds present in the environment than either would be alone. In Los Angeles the very irritating smogs that occur during temperature inversions, when the air is stagnant, result from the action of sunlight on a mixture of hydrocarbon and nitrogen oxide emissions from automobiles. The energy of the sunlight combines the emissions and produces nitrogen dioxide, which is far more irritating to eyes and throats than nitrogen oxide, and ozone which is harmful to health even in very low concentrations. On the other hand, two waste materials can interact so as to cancel the effect of each. Various salts

6

entering Lake Erie from human and animal wastes, detergents, and inorganic farm fertilizers act as nutrients that lead to "population explosions" of algae; after the algae die their decomposition uses up a great deal of the dissolved oxygen in parts of the lake and the resulting paucity of oxygen leads to a change in fish populations; the effects go on and on. Wastes that are (chemically) acid would counteract the basic salts. It has been suggested that if old automobiles were dumped in Lake Erie the acids from the metals might counteract the salts and reduce the algae growth – though I do not suggest that this would be either a feasible or a desirable way to attack the problem. Acid wastes discharged from steel mills probably prevent the development of a major algae problem in Burlington Bay at the western end of Lake Ontario.

When we are speaking of the harmfulness of various wastes, we must also remember that "one man's poison may be another man's meat." Water that is low in oxygen is desirable for some industrial processes and, as has been pointed out above, parts of Lake Erie are low in oxygen because of what most people consider to be pollution. And then there is always the problem of "how much?" A little fluorine is conducive to health, at least dental health, while too much is harmful.

It must be remembered that pollution of all kinds is always economically beneficial to some people to some extent. When a factory discharges untreated wastes into a water course, it is presumably adopting the cheapest way of disposing of its unwanted materials; pollution therefore keeps the price of the product lower than it would otherwise be; the factory owners therefore sell more product and make more money, and the consumers of the product buy it at a lower price than they would have to pay if pollution of the water course were avoided. Damages, or losses, in this case are suffered by those who want to use the water for other purposes that are inconsistent with its use for waste disposal, and what must be stressed is that those who suffer damages may well include consumers of the factory's product and the factory-owners themselves. It is only when the harm done by disposing of a particular waste in a particular way exceeds the benefits associated with the practice that a pollution problem

7

exists. To put the matter the other way around, the fact that there is always some gain associated with a particular instance of waste disposal is no proof at all that the practice is not socially harmful, for the damages done may exceed the benefits conferred.

The preceding paragraphs are intended to serve as a warning that when we classify waste materials and disposal practices on a scale of harmfulness we must remember that the classification is subject to the qualifying phrase "under normal conditions." The actual effects of a waste always depend on a long list of factors, including the amount of the waste, how it is disposed of, the condition of the environment into which it is injected, and the social and economic practices of the people that live in the area.

Old car cemeteries, glass and plastic litter, and slag heaps are disagreeable sights for most people, but (except perhaps for radioactive tailings) they do not constitute any significant health hazard nor do they normally involve any great economic costs. Slag and tailings piles, of course, occupy considerable areas of real estate which are thereby rendered useless for other purposes, but they are generally located in remote areas where land is not very valuable, so that the cost of using the land as a dump is not very high, at least in monetary terms; the disposal of solid wastes from large urban areas is, however, becoming an increasingly awkward and expensive proposition. Some waste materials adversely affect production processes and thus raise the costs of producing the goods concerned. As pollution in an area increases, factories that require particular qualities of air or water will either have to move to new (and presumably higher-cost) locations where pollution is not a problem, or remain where they are and bear the cost of purifying the water and air they use in their production processes. An increase of pollution levels also raises the cost of living for some groups of consumers, just as it raises the cost of production for some firms. Water pollution may increase the cost of recreation by leading people to build swimming pools or to travel further to unpolluted beaches; air pollution may lead to higher painting and cleaning costs, or higher commuting costs if the consumer solves his problem by moving away from heavily polluted areas.

8

Some wastes are known to be harmful to health, and the amount of harm usually associated with different "dosages" of a few of these noxious substances has been documented. Thus the danger of ingesting water with a given bacterial content resulting from sewage pollution is fairly well known, and public standards for drinking water and swimming are set accordingly. However, for most waste substances that are either known or suspected to be damaging to health, firm knowledge is lacking about the amount of harm done by given concentrations of these substances under particular environmental conditions. To the layman it appears that our scientific knowledge often peters out very quickly where pollution dangers and risks are concerned. We lack theoretical knowledge about the many possible interactions between different substances in the soil, the air, and the water; and, less excusably, we lack much easily obtainable information about the actual concentrations of many waste materials in particular environments and about the actual experience of different populations exposed to different types and amounts of wastes in their air and water. In Canada, the whole area of environmental health hazards seems to have been, and to continue to be, scandalously neglected.

Beyond the question of the health of men now living lies the question of the well-being of those yet unborn. Does one generation have any responsibility for future generations? Assuming that the well-being of future generations depends in part on the quality of the environment they inherit from us, should we be concerned about adverse effects of waste materials both on the physical environment and on non-human forms of life? Should we reckon the adverse effects of pesticide residues on birds, worms, and micro-organisms in the soil as a loss to be deducted from their advantageous effect on crop yields? Or is the present generation of one form of life – Man – the measure of everything? Should we care about the effects on future generations of our mistreatment of Lake Erie? Or should we do as we please and let future generations fend for themselves?

These are moral questions that do not admit of scientific answers, yet they demand the attention of all thinking people. So that the reader may be aware of the bias that underlies this book,

9

I should indicate my own attitude to the moral questions that are intertwined with pollution matters. I count myself among those who view with a mixture of distaste and alarm the growing number of instances of what might be called cavalier treatment of the environment, that is to say of human uses of the environment that are justified solely on the basis of net gain to those now living and in complete disregard of known or suspected harm done to other forms of life and to later generations of men. In terms of the preceding discussion of waste materials, one notes the dismaying coincidence that several of the wastes that are most durable and therefore accumulate through time – radioactive materials, insecticides and herbicides made from chlorinated hydrocarbons, and massive quantities of stable inorganic residues from detergents and artificial fertilizers – are also the materials that are most harmful to many non-human forms of life and, because they are cumulative through time, potentially harmful to future human populations. Some pollution, in my opinion, results from lack of knowledge, some from lack of imagination, and a great deal from sheer loutishness.

SUMMARY

The argument of this chapter, which has dealt with waste materials, may be summarized in six statements.

1 / Waste is the other side of the coin of production (or consumption).

2 / In terms of tons, it is equal to (or, in a growing society, slightly less than) the tonnage of natural materials that is used in the production process.

3 / The tonnage of natural materials used varies directly with the size of population and the standard of living, and inversely with the technological level of materials saving and materials recycling.

4 / Waste materials vary greatly in their durability.

5 / The social and economic costs of disposing of wastes and the hazards to health associated with different waste materials depend on a host of factors, including the nature of the material, its amount, the method of disposal, the condition of the environment

into which it is discarded, and many interactions between different waste materials after they have been discarded.

6 / People of a certain philosophical turn of mind define "harmful" wastes not only as wastes that are demonstrably harmful to the present human population but also as those that are demonstrably detrimental to non-human forms of life and that are potentially dangerous, directly or indirectly, to the well-being of future generations of humans.

The costs of waste disposal

It is only recently that pollution has been recognized as a social problem, and there is as yet no standard terminology in the field. In everyday conversations such phrases as "pollution damage," "the amount of pollution," and "pollution control" are used in a wide variety of senses. In this book I am going to make such frequent use of the word "pollution" that I dare not try to give it a precise meaning. Instead I shall use it in several senses: sometimes it will simply mean noxious wastes; sometimes it will refer to the damage done by those wastes; sometimes it will serve to suggest the vague concern people feel about the deleterious effects of wastes on the physical environment. But in the interests of clear communication certain phrases must be given a precise, technical meaning; I shall try to make sure that whenever I use one of these phrases it will always have the same meaning.

A BASIC VOCABULARY

Three of the technical terms I propose to employ are set forth in the following definitional equation:

waste disposal costs = pollution prevention costs
+ pollution costs.

The first term, waste disposal costs, I define as the total costs incurred by a society in disposing of its wastes. It is the sum of the costs that are paid to *prevent* wastes from causing damages plus the value of the costs incurred by *not* preventing wastes from causing damages. As we have seen in chapter I, there is always a net social cost of disposing of a waste, even though any particular method of disposing of it will confer benefits on some individuals. Accordingly, waste disposal costs, the sum of the net disposal costs for all wastes, are always positive. There is no way that society can avoid paying for waste disposal costs; but we can arrange things so that the bill is as small as possible.

By pollution prevention costs I shall mean the amount of money spent by public bodies or private parties to *prevent* some of the damaging or noxious effects of wastes. Examples are public expenditures on sewage treatment and garbage disposal, private expenditures by industries to process their wastes in some way before they are disposed of, and expenditures by individuals to reduce harmful exhausts from their automobiles. Private expenditures are sometimes made on private initiative; very often, however, they are made, willingly or unwillingly, in order to conform to the requirements of anti-pollution laws.

Pollution *prevention* costs are made to reduce the noxious effects of wastes *before* they are released into the environment; they therefore *prevent* pollution. Pollution costs are the money value of the damages caused by wastes *after* they are released into the environment; they are therefore the costs of the pollution that occurs. It is much more difficult to identify and measure pollution costs than pollution prevention costs. We shall distinguish three categories of pollution costs.

The first is public expenditures to prevent pollution damage (not to prevent pollution, but to prevent the damage that the pollution would involve if the expenditure were not made). The best example is the cost of providing water treatment facilities in order to render polluted water safe for drinking. Thus we call the cost of sewage treatment a pollution prevention cost, but the cost of water treatment a pollution cost.

The second category is expenditures made by private parties for the same purpose. These include the additional expenditures that people who live in polluted areas, as compared with those who don't, make for such things as the dry cleaning of clothes, the

13

painting of houses, medical bills, swimming pools, travel expenses to swim or fish in unpolluted waters, and air or water purification systems installed by industries whose production processes are adversely affected by impurities in the air or water.

The third category of pollution costs is the money *equivalent* of the reduction in welfare resulting from pollution damage that is *not* prevented. (Not only is the pollution not prevented; the damaging effects of the pollution are not prevented.) In some cases, as where families or firms avoid pollution by moving to non-polluted or less polluted areas, the cost are reflected in lower earnings or higher expenses. A man may move his family out of a city in order to avoid its air pollution, but continue to work in the city; he then incurs additional travel costs. Or he may move his residence and his job, probably at the sacrifice of earning a lower money income. Similarly, a firm may move to a new location in order to avoid air or water pollution, and as a result have to pay higher transport costs on its raw materials or its products. In other cases, the reduction in welfare involves no change in money transactions – no reduction in money earnings or increase in money expenditures – though the damage done can be *expressed* in terms of dollars. A smoggy day in Toronto makes no difference to my money income or my expenses – remember that I have already counted the extra cleaning and similar costs of smoggy days in my second category of damages – but I would nevertheless be willing to pay something to avoid the insult to my olfactory sense, my aesthetic sense, and my private sense of what constitutes public decency. Among welfare damages we must also include the value of pleasures foregone as well as the value of insults suffered. When local swimming becomes impossible because of pollution some people simply give up swimming; the loss in this case, however, is likely to be small since it is only the difference in the enjoyment they used to derive from swimming and the enjoyment they now derive from whatever "second-best" recreation they substitute for swimming. The loss in welfare as a result of un-prevented damage from unprevented pollution is none the less *real* because it is difficult, if not impossible, to measure in dollar terms. Indeed to many people it is this sort of damage that is the crux of "the pollution problem."

We can now summarize our technical vocabulary in two verbal equations. The first, already presented, is:

Waste disposal costs = pollution prevention costs
+ pollution costs.

The second is:

Pollution costs = public expenditures to avoid pollution
damage
+ private expenditures to avoid pollution
damage
+ the welfare damage of pollution (i.e., the
money equivalent of pollution damage that
is suffered rather than prevented).

So much for *concepts*. What are pollution *problems*? The over-all problem is to minimize waste disposal costs, given the amount of waste generated by a society. Amongst other things, minimization of total pollution costs involves choosing a correct balance between expenditures on preventing pollution, expenditures to avoid damage caused by pollution, and suffering the welfare damage of pollution. The last choice may sound more than a little paradoxical. What sense does it make to say that you can solve a pollution problem, even in part, by suffering pollution damage? We know, however, that to prevent either pollution or pollution damage costs money, and that no one is going to pay more to prevent damage than the money equivalent of the damage suffered. So it is most unlikely that we will pay to avoid all pollution damage. To the extent that we prefer to suffer the welfare damages caused by pollution rather than suffer the money costs of preventing them, we may reasonably say that some pollution is a good thing. The questions are always: "How much?" and "At what cost?" How clean you keep your car (and therefore how dirty a car you drive) depends on the cost of washing it (or having it washed) and on how much you dislike driving a dirty car. How clean you keep your environment (and therefore how much pollution you accept) depends on the cost of keeping it clean and on how much you dislike living in a dirty environment. Paying for cleaner air or water, however, involves very special difficulties that a person doesn't face when he decides how clean he will keep his car. The main purpose of this book is to begin to unravel some of

15

these difficulties. Before we tackle them head on, however, we must do some more thinking about pollution.

A DESIGNATED AREA

It seems impossible to identify any geographic region that would serve as a "natural" area for the study, or control, of pollution problems. Pollution is, of course, much worse in some areas than in others, yet no area is completely isolated from others since waste materials are more or less mobile, and different pollutants affect areas of vastly different sizes. Radioactive gases and airborne dust may contaminate the atmosphere of the whole world; different wastes discharged into a water system travel different distances, and thus affect larger or smaller areas before they are degraded or diluted below noxious concentrations; solid wastes in a township dump probably affect only the area within sight, or smell, of the garbage, though if the dump is a breeding ground for disease-bearing insects its noxious effects may be experienced over a considerably larger area.

But we are primarily interested in the number of *people*, not the number of *acres*, affected by pollution. Would areas of population concentration, then, serve as natural areas for the study of pollution? I think not. The reason is that people travel, so that they are exposed to, and concerned about, pollution in areas other than their area of residence. A great many Torontonians, for example, spend their vacations in Muskoka; their interest in the location and management of township dumps in Muskoka may very well be greater than their interest in Toronto dumps, since they spend much of their time in Muskoka during the warm weather. For some Torontonians, indeed, the air pollution in Los Angeles and Tokyo will be a matter of concern if they frequently find themselves in those cities. And there is even a sense in which water pollution in Mexico concerns Torontonians who have no plans to visit Mexico; after all they might travel there some day. These rather far-fetched examples aside, the point remains that as affluence increases and the cost of travel falls people become more mobile and therefore utilize, or might utilize, environments far removed from their area of residence. Both

16

pollutants and "pollutees" are mobile over greater or lesser distances, and it is this fact that prevents us from identifying any logically defensible "area" for the study of pollution.

We are therefore stuck with making an arbitrary choice. It seems wise, however, to choose a *political* area because we shall see later that pollution control is intimately tied up with political processes. Let us choose Ontario as our designated area. It is true that Ontario is too large a study area from the standpoint of the direct physical effects of many waste materials, and too small to encompass the area affected by others; it is too small to be the relevant environment for some people who live in Ontario and it is too large for others; it is not a fully satisfactory area from the standpoint of pollution control, because some pollution in Ontario originates outside the province, and some pollution originating in Ontario affects other areas. But in defence of the choice, it seems clear that the residents of Ontario do have the power to affect *significantly* the levels of soil, water, and air pollution within their province, and that this environment is the one they experience during most of their lives.

THE AMOUNT OF POLLUTION

In the first chapter we discussed the factors that determine the total amount of waste generated by a society. We also saw that Nature has a considerable capacity to absorb wastes without significantly adverse effects on the environment. The two main natural forces that help to reduce the noxious effects of wastes are the biodegradation of organic wastes, and the dilution of all wastes, both organic and inorganic. But the capacity of any particular body of water, air, or soil, to absorb punishment is strictly limited. Thus the more geographically concentrated is the disposal of wastes the greater the damage to the immediate environment and neighbouring environments – and, of course, the greater the damages suffered by human populations, since geographically concentrated waste disposal occurs, by and large, where humans are geographically concentrated. Thus pollution, in the sense of damages caused by wastes, tends to vary directly with the geographical concentration of human populations. If the

17

present population of Ontario were spread out evenly over the province I dare say that most, if not all, of our "pollution problems" would vanish.

Parenthetically, it may be noted that in most non-human forms of life the territorial imperative ensures a geographical spacing of individuals that avoids any natural pollution problem. More important is the fact that non-human populations utilize inorganic materials only where they occur. Man's technology enables him to make use of tremendous quantities of "inorganics" in small geographic areas and therefore to create problems of pollution by inorganic materials that do not occur in the plant and animal worlds. The same technology permits the geographical concentration of humans so that pollution by organic materials also becomes a problem for human populations that for practical purposes does not exist in non-human populations. It is urbanization, the concentration of large numbers of people in small areas, that "overloads" nature's disposal system in those areas, leaves unused much of the natural waste disposal capacity in lightly populated regions, and thus increases society's waste disposal costs.

This is an important point. While the amount of waste depends on the *size* of a population, the amount of noxious waste, or pollution, generated by a population of given size, depends to a large extent on the geographical *distribution* of the population. The more concentrated the population the greater the amount of pollution. The idea of concentration can be extended to include production processes and any other human activity, such as recreation, as well as people. The concentration of production into large-scale chicken-factories or hog-factories, and the concentration of recreation facilities in a small area, are all likely to lead to pollution problems. Concentration of people and production is a leading feature of modern life, and its inevitable accompaniment is a concentration of obnoxious wastes. Pollution is a product of our way of life.

The total amount of damage done by pollution, and especially the harm we have called welfare damage, depends in part on what effects people consider to be obnoxious and on how obnoxious they consider them to be. This problem – what people *perceive* as

18

pollution – is a very tricky one. Individuals differ greatly in their views about the gravity of any particular pollution situation. Bathers are annoyed when their favourite beach is closed to swimming because of water pollution; but boating enthusiasts using the same water may be quite unconcerned – indeed they may perceive the ban on swimmers as the removal of a hazard to navigation. Moreover people grow accustomed to, say, the odours in city air, and smells they considered obnoxious at one time will probably go unnoticed several years later. The difference in perception between generations is particularly important. Middle-aged people who remember Lake Erie as it used to be complain about the unpleasant swimming that the lake offers today; their children, who have learned to swim in private and public swimming pools, must often wonder why their elders get so upset about the condition of the lake.

Increased knowledge also has a lot to do with the perception of a pollution problem. No one thought about the contaminating effect of D D T on our environment twenty years ago, when the stuff was first used on a large scale, because no one knew what its effects on the environment would be. We still cannot be sure of the ultimate effects of its continued use, but we have learned enough about its dangers that we no longer use it as freely as we used to, and there are those who think that we should reduce its use (and the use of similar pesticides) even further, indeed to a tiny fraction of present sales. This subjective factor in the pollution equation is an important part of the pollution problem; it cannot be ignored, but it inevitably adds a certain amount of fuzziness to the concept. To a large extent, pollution is whatever people ("most people" or "the typical person") consider it to be. An amount of waste that would constitute a pollution problem in an Ontario city might very well be considered a normal part of life in New York or Tokyo or Calcutta today – or in an Ontario city thirty years hence. Any attempt to measure the amount of pollution runs into this problem, and it should be remembered that any estimate of pollution damage is subject to the qualifying phrase "given the present attitudes of the population."

If we take the attitudes of the population as given, the amount

19

of damage done by pollution depends on the cost of avoiding either the pollution or the damage caused by pollution, and this in turn depends on the level of "waste technology." Any technological development that makes it cheaper to process wastes into less harmful or less annoying forms either before or after they are injected into the environment will make it cheaper to buy cleaner air or purer water; and anything that becomes cheaper will be bought in larger amounts. One might hazard the guess that since pollution in many of its manifestations is a fairly recent phenomenon – the unwanted result of urbanization, the internal combustion engine, chlorinated hydrocarbons, detergents, and other modern products – our technology for dealing with it is still rather crude. Now that pollution is considered a serious social problem, we may expect more resources to be devoted to its "cure." With any luck at all, we should be able to invent more and better anti-pollution devices and waste-disposal systems in the not-too-distant future. Indeed, as will become clear presently, one purpose of this book is to outline an economic "invention" that I hope might serve as an effective anti-pollution device.

MINIMIZING WASTE DISPOSAL COSTS

Broadly speaking, pollution is a reflection of the amount of waste generated by a population and of the geographical concentration of population and production. In principle, then, pollution could be reduced either by cutting back on our production (and consumption) of goods and services or by reducing the degree of urbanization and spreading people more evenly over the landscape. In general these "solutions" to our pollution problems may be dismissed as absurd; most people would certainly consider the cure worse than the disease. It should be noted, though, that there is a lot of talk these days about "satellite" cities and the avoidance of large "conurbations." And there is some pressure, particularly from those responsible for garbage disposal, to reduce one particular component of our production – packaging. The cost of garbage disposal in large cities is being raised by our prodigal consumption of packages of all kinds, and also by the substitution

20

of plastics (which will neither rot nor burn) for paper in the packaging field. Packaging does seem to be overdone in many cases. Cigars, for example, are wrapped individually in cellophane, placed in cardboard packages (usually containing five cigars) that are themselves wrapped in cellophane, and the packages are then packaged in cardboard cartons that are *also* wrapped in cellophane. Surely there must be an easier and cheaper way to market fresh cigars!

Far from reducing our production or fleeing from cities, we seem intent, as everyone knows, on adopting policies that make our population, production, and urbanization grow apace. Inevitably, waste disposal costs grow even faster. Improvements in waste technology, the finding of cheaper ways of disposing of wastes or of reducing the damage they cause, help to dampen the rise in waste disposal costs. If we assume, however, that technology is constant, the only way of keeping total pollution damage at a minimum is to try to keep a proper balance between the three components of waste disposal costs: (1) pollution prevention costs, (2) public and private expenditures to avoid pollution damage, and (3) welfare damages suffered by individuals. By a "proper balance" I mean that if, for example, pollution costs become so high that it would be cheaper to prevent the pollution than to avoid its damage, it would obviously be wise to spend, say, $90 on a pollution prevention scheme in order to reduce private and public expenditures on pollution costs by, say, $100; such a move would reduce waste disposal costs by $10. In the remainder of this section we outline several of the main ways of dealing with pollution, and give a few examples of "expenditure switches" that may be, or may become, profitable moves in our never-ending war on wastes.

A
Pollution Prevention Measures and Costs

We have already noted two examples of pollution prevention costs – public expenditures on sewage treatment and the private expenditures that automobile owners are now being required to make on various devices to reduce noxious exhausts. It has been decided that, with the increasing number of automobiles spewing noxious wastes into a fixed amount of air, welfare damages have become so high that it is now profitable to reduce them by spending more money on pollution prevention. The increasing amount of money being spent on sewage treatment plans is similarly based on the judgment that, with increasing populations discharging their wastes into water sysems, it has become more economical to spend money on pollution prevention than to suffer the rising damage avoidance costs and welfare damages resulting from increased pollution. Sewage treatment plants currently in use, however, are of limited usefulness. In particular they are incapable of removing various nutrient salts from organic wastes, and in some lakes, notably Lake Erie, these nutrients have led to severe algae problems. Algae problems in turn impose damage avoidance costs in various forms – expenses for swimming pools, extra expenses for vacations in areas further from home, and perhaps additional expenses of treating the water for domestic or industrial uses – and also welfare damages, largely in the form of the stink of rotting algae that are washed up along the shores. As these costs mount, it may very well become economical to reduce the inflow of nitrates and phosphates into the water system by expenditures on pollution prevention – more complete treatment of sewage in "tertiary" sewage plants and more careful farming practices to prevent undue amounts of farm fertilizers and farm animal wastes from entering water courses.

Other examples of pollution prevention costs range from the expenditures incurred to keep dangerous radioactive wastes out of the environment (it is intended to store some of them for a thousand years or more, this being the length of time required for some radioactive isotopes to work off most of their malignant energy) to the costs of disposing of household and other types of garbage. The garbage disposal case again illustrates the economic considerations that permeate pollution problems. In Toronto, for example, garbage has for long been dumped in some specified tract of unused land and carefully covered with earth, a technique that goes by the quaint name of "sanitary land fill." Suitable sites for these sanitary dumps are almost exhausted, and the city must soon decide whether to pay the cost of transporting its garbage to new sites some distance away, or to burn, in huge incinerators, as much of it as can be burned. In either case we may have to retreat a bit on this part of the pollution front; the new techniques are certain to be more expensive than the old, and both threaten to add something to air pollution, either in the form of truck exhausts or incinerator emissions.

Mention of the possibility of transporting garbage away from people leads to the observation that, in principle, all wastes could be handled in a similar fashion. Indeed if all wastes were hauled away from cities and spread evenly over the countryside the result, from the standpoint of pollution, would be much the same as if people and production were spread evenly over the landscape. Waste disposal costs would be reduced somewhat because the capacity of natural forces to dispose of wastes would be more fully utilized; but the costs of the enormous amount of transportation that would be involved (not to mention the increased wastes in the form of exhaust fumes that would thereby be generated) mean that the loss would be much greater than the gain.

There is, however, one example of this "transport and disperse" technique that might prove to be profitable. Modern chicken-factories turn out not only broilers but small mountains of guano. The guano makes good fertilizer – the same is true, of course, of the manure from hog-factories and the body wastes of human populations – but the amounts that have to be disposed of far exceed the amounts that farmland within economic trucking distance of the factory can absorb. It also seems to be the case that it is profitable to ship commercial inorganic fertilizers much further than it is profitable to ship guano, probably because they are more concentrated. The chicken-farmer, or the city sewage official, thus either has to discard the wastes for which he is responsible into a water system (and tougher water pollution laws are beginning to reduce this possibility) or pile them up on some piece of land set aside for the purpose and take measures to prevent their becoming insults to the eye and nose. Perhaps techniques could be found to enrich these wastes so that they could, like manufactured fertilizers, be economically shipped over

23

greater distances. Even if this were not possible, it might be wise to subsidize their transportation so that they could compete over greater areas with artificial fertilizers. The fertilizer companies would probably not like the idea, but from the standpoint of the whole population, the subsidy would be costless so long as its amount did not exceed the reduction in expenditures on manufactured fertilizers. An even larger subsidy could indeed be justified as a pollution prevention cost, so long as it did not exceed the reduction in expenditures on manufactured fertilizers plus the present costs of disposing of the guano. "Transport and dispersal" seems to make good ecological sense in the case of organic wastes from modern farm-factories; it recreates, in a sense, the situation that used to exist when chickens and hogs and cattle were spread more evenly over the countryside than they now are and when their wastes helped to preserve soil fertility rather than add to pollution problems. So far as I know, however, the economics of this suggested switch in disposal practices have not been investigated, and only after a full accounting of all costs of both the present and the proposed system would we be in a position to know whether it would be profitable to make the change.

It might be thought that instead of dumping wastes evenly over the countryside, the simpler and much cheaper procedure of dumping wastes from populated centres at a few sites in unpopulated regions should be considered an anti-pollution measure, either a pollution prevention cost or a pollution cost. One consequence, however, of choosing Ontario as our "designated area" is that everyone in Ontario must be thought to be as concerned about pollution in one part of the province as in another, regardless of its population. Thus if all the obnoxious waste generated in Toronto could somehow be dumped in Muskoka or around James Bay, we would, collectively, be no better off; we would not have reduced pollution, but only relocated it, rather like a careless maid who makes the dust fly from the living room to the dining room.

Obnoxious industrial wastes can often be reduced in quantity by a change in production processes, especially by the introduction

24

of equipment to remove useful materials from waste waters or gases. Except in a few cases where entrepreneurs find, to their surprise, that waste-reducing practices are actually profitable, the new processes will be more expensive than the old; if they are introduced the additional cost of production must be reckoned as a pollution prevention cost.

B

Damage Avoidance Measures and Costs

Water treatment plants provide the main example of public expenditures to avoid pollution damage, as distinct from expenditures to prevent pollution. Other examples are municipal expenditures on street cleaning and swimming pools, and provincial expenditures on removing highway litter. In the latter case it might be far cheaper to raise fines for littering so that the litter would not be generated in the first place than to pay for its removal after it has occurred.

Most examples of damage avoidance costs, however, take the form of private expenditures. We have already mentioned several of the costs people incur in order to offset damages caused by air or water pollution – increased cleaning bills of all kinds, more expensive recreation, and probably increased medical expenses. So long as all areas are not equally polluted, moving away from heavily polluted areas is the simplest and usually the most effective way for individual families or firms to avoid pollution damage. The solution is not costless – no solution ever is – since the new location will involve either lower money earnings or higher money costs, or a sacrifice of some of the non-monetary advantages of living in a big city. Nevertheless, the cost may not be very high, and especially for families or firms that find big city pollution particularly objectionable, it will probably be well worth paying. Perhaps more people than we think have already chosen this solution. People who move to a city, or do not move out of it, may be those whose aversion to pollution is not very great; if this is so, it might help to explain why city electorates seem so reluctant to elect politicians who will spend more money for pollution control.

25

C

Welfare Damages

As population grows, urbanization increases; and as the standard of living moves up, pollution soars. Traditional pollution prevention or damage avoidance techniques prove increasingly inadequate to cope with the situation, and the result is that the welfare damage component of society's waste disposal costs rises sharply. Welfare damage – pollution damage that is not offset by damage avoidance costs – is increased by growing suspicions that air pollution is harmful to health in many insidious ways, by more dangerous smogs, by growing pollution of established vacation areas, by increasing garbage litter throughout the countryside, by experience of algae problems, by growing awareness of the dangers of chemical pesticides, and by the vague awareness that human mistreatment of the environment might just possibly cause changes in the ecological balance that would gravely affect the well-being of our grandchildren, if not our children. Growing pressure is likely to develop to transform welfare damages into pollution prevention or damage avoidance costs. At some point, we simply *must* decide how much we will spend in order to improve the quality of our environment.

But we will also, if we are wise, find out *how* to spend whatever we decide to spend so as to get the greatest value for our money. That is where economic analysis can prove particularly valuable. Not that economics has all the answers; far from it! But we need any help we can get in meeting the challenge of pollution.

26

Simple problems, simple solutions

Economists have found that it is usually very helpful to attack complex problems like pollution by assuming away all their complexities and then solving the artificially simplified problems that remain. The value of the technique lies *not* in the answer to the artificial problem (ask an artificial question and you'll get an artificial answer) but in the making of the assumptions that allow us to solve it, for these assumptions help us to identify exactly what features of the original problem make it complex and difficult. And it is only when we know exactly what the difficulties are that we can begin to zero in on them. Let us, then, begin our study of the economics of pollution with a very simple problem, taking great care to note exactly *why* it is so simple.

THE FIRST PROBLEM

Imagine a city of 100,000 voters located on the shore of a small lake. There is only one pollutant that enters the lake, human wastes from the city's population. Pollution of the lake has been gradually increasing through the years because the inflow of wastes exceeds the capacity of the lake to degrade or dilute them sufficiently to prevent the quality of the water from deteriorating; until very recently, however, this fact has been unknown, no one having taken the trouble to keep track of what was going on in the lake. At the time our analysis begins the lake water has just been declared dangerous for drinking purposes, and the citizens find themselves face-to-face with a pollution problem.

Now for the assumptions that will allow us to solve the problem:

1 / There is only one pollutant.

2 / There is only one group of polluters; and each member of the group pollutes to exactly the same extent as every other member.

3 / Each member of the group takes exactly the same view of the medical warning as every other member.

4 / Each member would be willing to pay up to $10 a year, but no more, in order to avoid the risk involved in drinking water that might be harmful to his health.

5 / The risk can be avoided in two, and only two, ways: each family can boil its own drinking water at a cost per voter of $6 per year, or $600,000 for the entire population; a city water treatment plant can be provided at a cost of $250,000 per year, or $2.50 per voter.

6 / Everyone expects that within the next ten years the lake will be declared unfit for swimming. Each would be willing to pay a dollar per year from now on to avoid that eventuality. They discover, though, that the cheapest way of avoiding it would be to pay for sewage treatment, which would cost each of them $3.00 a year. (The sewage treatment plant would not solve the drinking water problem: see assumption 5.)

7 / No one expects further damage, except to swimming, from continued pollution of the lake.

The answer is easy, isn't it? The citizens vote to build a water treatment plant, and that's that! (The swimming problem, of course, is still hanging over their heads, but it is not worth doing anything about it right now, so they forget it.) Even though the answer is self-evident, it will be well to construct a simple table – which we shall call a benefit-cost table – to make sure we know *how* we got the answer. Looking at Row 4 of the table we see that the policy of building the water treatment plant is indeed the best solution to our problem because the net benefit of this policy is greater than that of any other (including doing nothing about

Benefit-Cost Table I

1 POLICY	water treatment plant	boil water	water and sewage plants	do nothing
2 GROSS BENEFIT value of damage avoided	(health) $10.00	(health) $10.00	(health and swimming) $11.00	$ 0.00
3 COST OF AVOIDING DAMAGE	$ 2.50	$ 6.00	$ 5.50	$ 0.00
4 NET BENEFIT (item 2 − 3)	$ 7.50	$ 4.00	$ 5.50	$ 0.00
5 WELFARE DAMAGE REMAINING	(swimming) $ 1.00	(swimming) $ 1.00	$ 0.00	(health and swimming) $11.00
6 WASTE DISPOSAL COSTS (item 3 + 5)	$ 3.50	$ 7.00	$ 5.50	$11.00

the problem). Row 6 gives us a further check on our reasoning because it shows that our society's total waste disposal costs are minimized by building the water treatment plant. We have therefore found the proper balance between the various components of waste disposal cost. (We could get rid of the welfare damage of $1 by building a sewage plant, thereby raising our total waste disposal costs from $3.50 to $5.50; but it does not make sense to raise costs by $2 in order to get rid of damage valued at $1.)

We must now ask *why* the answer to our first problem was so easy? The first reason is the so-called people we have invented. They are certainly a dull lot, because they are all exactly the same (assumptions 2, 3, 4, 6, and 7). It is almost as if we were dealing with one individual instead of a society of individuals because the social problem is simply 100,000 times any individual's problem; and the solution to the social problem is simply 100,000 times any individual's solution to his own problem. There really wouldn't be any *social* problem if we all thought alike – and the essence of the pollution problem in the real world is that most people don't seem to take it as seriously as we do! Our "people" are also pretty ignorant (assumption 7), since damage to other uses of the lake will certainly become manifest if pollution continues unabated. Some of us may be willing to say, moreover, that they are a pretty insensitive lot because they know they are going to ruin the lake for swimming and really ought to be willing to pay for that sewage treatment plant; but then *we* don't all think alike, especially when it comes to opinions about what people ought to do.

The second reason why the answer is so easy is that everything about the problem is so cut-and-dried; all the benefits (that is, all the damages avoided) and all the costs of every possible line of action are known, are measurable in dollars, and have in fact been measured and presented to the citizens in dollar amounts. Choice is simple when everything is cut-and-dried, and everything is cut-and-dried when all aspects of a problem can be exactly measured in some common unit, such as dollars. There are two reasons why our artificial problem can be expressed in exact dollar terms; we shall label them reasons 2A and 2B. Reason 2A is that everyone knows exactly what his welfare damage would be if he did nothing about his pollution problem (assumptions 4 and 6). Reason 2B is that the only possible solutions are "all-or-nothing" solutions (assumptions 5 and 6); we shall look at a problem in a moment where it is possible to buy different degrees of protection from risk for different prices.

The third reason why the answer to our initial problem is so easy is that the technical problem is very simple, mainly because there is only one pollutant (assumption 1). If there were ten pollutants it would not always be possible to calculate the damage done by each, and then add them up to get the total damage. As we have seen, pollutants often interact, sometimes to escalate their individual damages, sometimes to reduce them; pollutant damages are not always additive, as the mathematicians would say. When two pollutants, x and y, escalate, and you pay to get rid of one of them, say x, you get a bonus because y is much less damaging by itself than it was with x. On the other hand, if x and y are off-setting, getting rid of x will increase the damage caused by y. (In the early days of chemical insecticides, D D T was used to kill coddling moths in apple orchards; it did not, however, kill

30

mites, whose numbers had formerly been kept in check by coddling moths, who ate them; the result was that after the moths were killed the mites had a field day, and orchardists suffered more mite damage than they would have believed possible.) In this book, however, I am going to ignore the non-additive aspects of pollution problems. They only complicate the calculation of the damage figures that the economic analysis of pollution uses; they do not affect the analysis itself.

There is, finally, a fourth artificiality that must be mentioned, a fourth reason why our first problem is so easy to solve. We considered only one city, one area (assumption 2), so we didn't have to worry about how pollution in one area by one group might affect the well-being of other groups in other areas. Yet we agreed earlier that as citizens of Ontario we have to be concerned about pollution in many different areas. When we remove the "one-area" artificiality from our analysis, complications come flooding in.

In summary, we have *made* our problem easy – even ridiculously easy – by imposing on it a number of simplifications. Two of these are unimportant because they only simplify technical aspects of the problem: we shall not discuss further the assumption of only one pollutant (reason 3) because in the real world engineers could in most cases provide good data on the effects of more than one pollutant; and we shall show in a moment how we can dispense with the reason 2B simplification of "all-or-nothing" solutions. But all of our other carefully contrived constraints on our simplified "pollution" problem touch on essentials; they not only simplify actual pollution problems – they also distort them. In the following chapter we shall discuss a more realistic, and therefore very much more complicated, situation. For the moment, however, consider a second artificial pollution problem in which we do without the 2B simplification of our first problem.

THE SECOND PROBLEM

The 2B simplification made pollution problems "all-or-nothing" propositions – they were either solved or not solved. Sometimes, perhaps, that is the way things are; but usually we have to choose

31

among different "degrees of solution" to pollution problems. (The question is not whether I will keep my car clean or dirty, but how clean – or dirty – I will keep it.) Let us therefore change the character of our first problem so that "partial solutions" as well as "all-or-nothing" solutions become possible, and see whether we can find an answer to this more complicated problem.

The voters are now told by the City Engineer that he can provide them with a treatment plant that will remove all of the noxious wastes from drinking water for a cost of $5 per voter per year; or one that will remove 75 per cent of the wastes at a cost of $3.50, or one that will remove half the wastes (and thus prevent half the damage) for $2 per voter per year. Each voter now asks himself how much he would pay to avoid all the damage, three-quarters of the damage, and half the damage that he expects to suffer if nothing is done about the condition of the city drinking water. Since all the voters are exactly the same, we might as well stick with the City Engineer, who is of course a voter. He says to himself: "If nothing is done I expect to be sick one day a year, and I'll pay $10 to avoid a day's sickness; thus a treatment plant that was 100 per cent effective would be worth $10 to me. If only half the wastes are removed I expect I would be sick only about three days in ten years; I think the harm done by wastes goes up faster than the amount of wastes I imbibe, so that as the dosage doubles the harm more than doubles; so if I halve the dosage I'll cut down my sickness by more than 50 per cent, by more than half a day a year; three-tenths of a day a year sounds about right, so a treatment plant that would cut my losses from $10 to $3 a year would be worth $7 a year to me. Similarly, if a plant that is 75 per cent effective cut my sickness to one day in ten years, it would save me $9 a year in sickness costs; that's what the plant would be worth to me."

The Engineer then recalls that it would cost him $6 a year to boil his drinking water, and with the help of his slide rule he quickly calculates that if he boiled half his water (or all of his water half the time) the cost would be $3, and that to boil it 75 per cent of the time would cost $4.50. The Engineer then thinks a moment and decides that he will refer to "the value of sickness

32

Benefit-Cost Table II
Two Measures with Varying Degrees of Treatment

	100%	75%	50%	0%
1 DEGREE OF TREATMENT	100%	75%	50%	0%
2 GROSS BENEFIT value of sickness avoided	$10.00	$9.00	$7.00	$ 0.00
3 COST OF AVOIDING DAMAGE boiling city plant	$ 6.00 $ 5.00	$4.50 $3.50	$3.00 $2.00	$ 0.00 $ 0.00
4 NET BENEFIT (item 2 − 3) boiling city plant	$ 4.00 $ 5.00	$4.50 $5.50	$4.00 $5.00	$ 0.00 $ 0.00
5 WELFARE DAMAGE REMAINING* boiling city plant	$ 0.00 $ 0.00	$1.00 $1.00	$3.00 $3.00	$10.00 $10.00
6 WASTE DISPOSAL COSTS (item 3 + 5) boiling city plant	$ 6.00 $ 5.00	$5.50 $4.50	$6.00 $5.00	$10.00 $10.00

*Value of sickness not avoided, i.e., $10 minus item 2

avoided" as the "gross benefit" of each scheme, and that he will describe the costs of boiling water or paying for a water treatment plant as "damage avoidance costs." The "net benefit" of any given scheme he defines as gross benefit less damage avoidance costs.

This is all the information the Engineer needs to solve his problem, but, being an engineer, he wants to check his reasoning by working out the problem another way. To do so, he defines two more magnitudes. Some of the schemes he is considering do not avoid all the pollution damage, and the "psychic" cost to him of the risk of sickness that he does *not* avoid by adopting any particular scheme he decides to call "welfare damage." The total cost to him of dealing with the pollution problem is therefore the sum of his damage avoidance costs (paid in money) plus the value of the welfare damage (which he simply grins and bears); this sum he labels "waste disposal costs." He is now ready to put all his findings into a table.

The effort of preparing the table has obviously been worthwhile, because the solution now leaps to the eye. We choose the scheme with the *highest* net benefit, which turns out to be the 75 per cent effective city treatment plant. By looking at Row 6 we see that this plan results in the lowest total cost of disposing of the wastes – which is, of course, as it should be. By adopting the 75 per cent treatment plant, total waste disposal costs have been reduced from $10 to $4.50; a profit or net benefit of $5.50 has been made on the deal. (On either the 100 per cent treatment plant or the 50 per cent treatment plant the profit would only be $5.)

It is obvious, of course, that the answer we get depends entirely on the numbers we have assumed. If, for example, the cost of providing a 100 per cent treatment plant had been $4 instead of $5, and all the other numbers had remained the same, the net benefit of this policy would have been $6, waste disposal costs would have been reduced to $4, and our best choice would have been the 100 per cent treatment plant. Or if the costs of all treatment plants were $2 higher than the figures we have chosen, the best solution would turn out to be to boil all drinking water. You can play around with the figures all you like, and rig them in such a way as to get seven different answers – one for each active policy and one for a policy of doing nothing. (If the costs of avoiding damage by all of the six "positive" policies is greater than $10, then the best policy is to do nothing.)

You can imagine that tables of the type illustrated above could become very complicated if there were, say, half a dozen different types of damage to be considered and seven or eight different ways of either avoiding it (damage avoidance schemes) or preventing the pollution that occasions it (pollution control schemes), each of which would probably help to control different proportions of different types of damage. Benefit-cost calculations, as they are called, of actual pollution problems might easily become so complex, especially if we were to include the Reason-3 difficulties occasioned by interacting pollutants, that we might very well want

34

to buy a little bit of high-speed computer time in order to work them all out and arrive at the net benefit figures.

What is important, though, is that they always *can* be worked out – provided, of course, that we remain within our very artificial world where Reason-1 assumptions protect us from real people and let us deal with very dull "creatures," all of whom are exactly alike; where 2A assumptions ensure that we have all the information we need to solve the problem; and where the Reason-4 artificiality means that we are dealing with a one-community world.

A CONCLUDING CAVEAT

A major purpose of this chapter has been to illustrate a certain way of thinking about pollution problems. The technique has been to balance the benefits against the costs of different policies (including a policy of doing nothing) in order to find the most profitable policy. This "cost-benefit-profit" apparatus is simply a formal description of the process of making a choice. Every day everyone makes choices, i.e., decides to do this rather than that. In order to get the benefit of doing "this" one must incur a cost, even if the cost consists only of giving up the pleasure (or profit) associated with doing "that." Every line of action (including "zero-action" or inaction) involves both costs and benefits, and sensible behaviour consists of nothing more than adopting that line of action that yields the greatest profit, i.e., the greatest excess of benefit over costs.

In the present chapter the benefit-cost apparatus has provided us with an orderly way of thinking about our simplified problems; and it will serve the same function in the discussion of *any* problem of choice, no matter how complicated. A guide to straight thinking is not, of course, a dramatic break-through in the pollution field, but it does constitute an essential base for any rational attack on the problem. If you look carefully at many proposed "solutions" to pollution problems you will be astonished to find how often authors forget that every benefit has its cost and how often, in their enthusiasm, they argue as if the great benefits of their pet anti-pollution schemes could be obtained at no cost.

In the simplified problems dealt with in the present chapter, benefit-cost analysis has perhaps seemed to be much more than a guide to straight thinking; it may have seemed, indeed, to be a magic machine that actually solved problems for us. But do not be deceived! The problems were *not* solved by the benefit-cost machine; they were solved because we took great pains to adopt assumptions that *made* them solvable. A brief discussion of some of the complications involved in choices about real-world pollution problems will serve to suggest why the actual making of decisions about such problems cannot be reduced to a mechanical, numerical procedure.

We all know that a personal decision about a major matter – say the choice of a career – is apt to have an important effect on other aspects of our lives: where we live, our opportunities for travel and recreation, our income level, and perhaps even on our choice of mate. Everyone faced with a major personal decision is aware of such interrelations and in trying to make a wise choice from among the different possibilities open to him no doubt makes a sort of mental cost-benefit analysis; but no one, surely, would maintain that he could actually draw up a numerical benefit-cost table that would show him how to live happily forever afterwards! The main reason why we would have little faith in numerical "measurements" of benefits and costs that relate to major decisions is, I think, the realization that what we decide today may well affect our lives so much that ten years from now we will probably look on things very differently from how we look on them today. The result is that we will probably never know whether we have made the "best" choice available to us. Twenty years ago I could have decided to be an architect instead of an economist; I have no way of knowing now whether I chose wisely or not. Indeed the very concept of choice becomes tenuous when we deal with areas of life in which there are normally very few second chances (so that we have little opportunity to learn by a trial-and-error process) and where what we do today has a strong effect not only on what we are likely to be doing tomorrow, but also on what sort of person we are likely to *be* tomorrow. It is notable, though, that despite the uncertain outcome of such

36

choices – or, somewhat perversely perhaps, because of the uncertainty – we always make special efforts to come to "wise"
decisions about them. In the choice of career, for example, we
observe the life-style of people in different occupations and seek
to assess our own strong points and weak points in an attempt to
make as fruitful a combination as possible between our capabili-
ties, our desires, and the apparent demands and rewards of different career lines. And despite the fact that very few of us would
pretend to be able to make (or to have made) a scientific decision
about choice of career, most of us probably make a reasonably
sensible choice, though we can never be sure that we couldn't have
done better.

Major social decisions, such as the choice of a pollution policy,
are analogous to major individual decisions. We know that what
we do about pollution (or don't do about it) is going to affect
many other aspects of our lives – the cost of living, where we live,
where we work, what we do for recreation, and so on, and on.
Society is bound to adjust in scores of ways to whatever policy
about pollution is adopted, and in order to list all the benefits and
costs of a given policy we must try to imagine what the new
situation would be like – just as, in deciding on a career, we have
to try to imagine what it would be like to be a doctor, or a farmer,
or an economist. To some extent the social decision may be more
amenable than the individual decision to objective calculation;
social scientists, at least, believe that it is easier to forecast the
average reaction of a large number of interacting individuals to
a change in circumstance than it is to forecast the reaction of any
particular individual. We therefore have some basis for "imagining" the changes in society that would result from the adoption
of any given rule about pollution. As a practical matter, of course,
it is much easier to predict the direction of a change than its
magnitude, so that social science predictions in numerical form
are usually very inaccurate. Moreover, when we make decisions
that will in all likelihood affect the well-being of people twenty
years hence – including those yet unborn – we cannot assume that
"people then" (including those of us who survive) will have the
same attitudes about pollution as those held by "people now." It

then becomes impossible in principle, as well as in practice, to choose the best pollution policy for all concerned – unless someone is willing to gaze into his crystal ball and forecast what people's attitudes will be twenty years from now. But again, as in the corresponding individual decision, gross uncertainty about the outcome of a decision should serve as a challenge to make it carefully and in the light of as much knowledge as possible, not as an excuse for pretending that questions that cannot be answered scientifically are not worth worrying about.

Pollution problems are *social* problems, and decisions about them must result from some process of "social choice." I have so far avoided the problem of whether there actually exists a unique, best solution to any "social choice," and for the most part I shall go right on avoiding it because it is probably insoluble. The problem is simply that if ten people have to make a joint decision about something – say a pollution question – there will be at least ten views about what decision should be made. How, then, can individual preferences be combined to produce a "social preference"? What is really at issue here is a set of age-old questions: What is a society? How should it be governed? How can it make decisions that reflect the general will? The long search for utopia has not so far yielded completely logical answers to these questions; actual governments are only more or less workable institutions, not fully logical creations. If we could find some benefit-cost analysis that would always produce the best answer to social questions we would, in fact, have found utopia.

Which is unlikely. We should be content that benefit-cost analysis provides us with an orderly way of thinking about pollution problems, and not try to make it into an all-purpose decision-making machine. Indeed to most problems in social choice, as to the most important of all social questions – how should we govern ourselves? – there is probably no one best solution. Social decisions lead to social change. But who knows the way to utopia?

Actual problems, actual solutions

ECONOMIC INSOLUBILITIES

Can economic analysis be used to pick out the best anti-pollution policy from among, say, a score of policies that have been put forward to deal with some pollution problem in the real world where people differ, where information is incomplete, and where we must think of both problem and policy not only from the standpoint of those who live in the area affected, but also from the standpoint of everyone else living in Ontario? Or can economic analysis be used to *devise* a policy that would be better than any of those actually proposed? The answer to both questions, in my opinion, is a simple "no." Once we discard the three simplifications that allowed us to solve the artificial problems of chapter III, the economist is quite unable to draw up a neat table showing all benefits and all costs of all anti-pollution policies that are proposed (or that might be proposed); he is therefore quite unable to say that one policy is demonstrably superior to all others.

But this is not to say that economics is altogether useless in dealing with real-world pollution problems. There is one sense in which economic analysis can almost always be very helpful. I have so far been using the word "policy" to mean "what is to be done." But even when we have decided what we will do, we still have to decide how we will do it. In resolving this second question of means to achieve some given end of policy, economic analysis is, I think, indispensable; indeed I think that it can almost always lead us to the best answer. But more of this in chapter VI.

At the moment, the subject is humility. We argued in a rather abstract way at the end of the last chapter that it may very well be that no one "optimum" solution to social problems exists. Here we shall be more down to earth. Even if social problems are logically insoluble, something is always done about them, and the practical question is always whether we can do something *better* about them than we are now doing. In the past our policy about waste disposal has been (to exaggerate somewhat): "Let everyone dispose of his wastes in whatever way is cheapest for *him*." A rising tide of complaints about pollution indicates widespread dissatisfaction with this policy, and a desire to replace it with a new policy that will take into account not only the costs of waste disposal to the individual who discards the waste, but also the costs *to all other individuals*. There are, of course, an infinite number of policies of this sort and some of them will no doubt be adopted. We do not lack for solutions; what we do lack is any way of deciding – not in principle, but in practice – whether one solution is any better than another. Why do we find ourselves "hung up" on this question?

It is the lack of information that is the crux of the matter. Information may be lacking for two reasons: because it is not available, though obtainable; and because it is unobtainable, and therefore never will be available. Information lacks of the first variety may be very unfortunate in practice; but they are not very important to theorists who are mainly interested in how a problem might be solved in principle. If the information could be obtained the theorist has no qualm whatever about assuming a figure in order to get his solution. (In chapter III, I assumed figures for the costs of boiling water and for the costs of water purification plants; even though I have not the slightest idea what these costs are, I know that they are either known already or could be found out.) But it seems somehow "unfair" to assume a figure that is not only not available but also unobtainable. (Not only do I not know what the subjective value of a day's sickness to the City Engineer

ACTUAL PROBLEMS, ACTUAL SOLUTIONS

in chapter III is; I haven't the faintest idea how I could find out what the welfare damage of a day's sickness suffered by a real city engineer is. As we shall see presently, we are not likely to find out simply by asking him – and several millions of other people.)

The amount of information relevant to pollution problems that is unavailable, though obtainable, is staggering. We do not know enough about currents in Lake Ontario to know whether, on average, Oshawa pollutes Toronto or *vice versa*. Very likely, Oshawa pollutes Toronto part of the time; Toronto pollutes Oshawa part of the time; and neither pollutes the other part of the time. From 1884 to 1893 three people who were curious about the currents in the Toronto area of Lake Ontario made some experiments to find out what they were; in 1966, the Great Lakes Institute at the University of Toronto became curious about the same question and made further experiments to try to find out more about it. The results of both sets of experiments were broadly similar; we now know that the direction and force of currents in the Toronto area of Lake Ontario vary both daily and seasonally. But since the currents sometimes flow from west to east and sometimes from east to west, we still do not know whether, on average, Toronto pollutes Oshawa or Oshawa pollutes Toronto. On a river, of course, we know that a downstream community never pollutes its upstream neighbour; but it is always very difficult – and therefore expensive – to find out by exactly how much an upstream community pollutes the water of a downstream community. When we deal with lakes, however – and in Canada more people live on or near lakes than on rivers – we do not even know who pollutes whom, let alone by how much. Instead of going to great expense to try to find out who pollutes whom by how much, whether on lakes or rivers, it seems much more sensible to take the view that we all pollute each other more or less equally, and to treat our pollution problems as common problems, i.e., social problems.

About the relationships between various types of air pollution and health we really know very little. We can find out, if we are willing to experiment on human beings, how much sulphur dioxide inhaled over a given period of time, is required to induce

41

coughing in a quarter, or a half, or 80 per cent of the people being studied. Knowing this, we still would not know, however, whether much smaller concentrations inhaled over much longer periods of time, say ten to fifteen years, would be harmless. Moreover, the harm done might very well depend on what other substances were present in the air along with the sulphur dioxide.

All of this objective, scientific information is, however, obtainable. It would of course, cost money, time, and effort to obtain it, and there is always a great deal of potential information that we will not have because its value to us seems too small to justify the cost of getting it. It is very important, however, that we fill in the most obvious gaps in our scientific knowledge about pollution as quickly as possible so that we may have a much better understanding than we now have of how our waste disposal practices affect the environment and how changes in the environment affect us.

But even if we had all the scientific data we wanted, we still would not be able to construct a fully reliable benefit-cost table for an actual pollution problem. In the artificial problem discussed in the previous chapter, good scientific information allowed the City Engineer to make an accurate estimate, rather than a rough guess, of how many days' sickness were associated with different degrees of water treatment. This physical measure of damage then had to be translated into dollar terms before the benefit-cost table could be drawn up. And there was no difficulty in doing so; all we had to do was ask the Engineer how much he would pay to avoid a day's sickness. We assumed that he knew *exactly* (ten dollars—no more, no less) and that he answered us honestly; and since we knew that the Engineer was exactly like everyone else, we knew that everyone else would suffer the same amount of sickness as the Engineer for a given degree of water treatment, and that everyone else would value the avoidance of a day's sickness at exactly ten dollars – no more, no less.

But try asking your friends how much they would pay to avoid a day's sickness. The answers, if you can get them, will vary tremendously. Some will want to know which day you are talking about – Sunday or Tuesday; some will want to know if they are going to be sick enough to require a doctor or drugs, or just sick enough to have to stay away from work for a day or two. Salaried people are not likely to lose any pay for a couple of days' absence from work; lawyers, dentists, or surgeons may lose quite a bit of income; and hourly paid workers will know exactly how much pay they will lose. Or ask a group of Torontonians how much they would pay to be able to swim along the Toronto lake shore; or how much they would pay to prevent the sulphur dioxide content of Toronto air from exceeding one part per million for, say, 358 days per year, or two parts per million at *any* time. Or how much they would pay to subsidize the shipment of guano away from a chicken-factory on the outskirts of Cornwall. People really can't answer these questions, though to the last one there would probably be many snorts of "Not a damn cent!"

I think you will agree that there is no hope of estimating the welfare damages of pollution by simply asking people what they would pay to avoid them. I don't think they know. And if they do know, some suspicious social scientists have suggested that they won't give an honest answer anyway. They may greatly overstate the damage they suffer, and thus magnify the problem, if they think that somebody else ("big industry," "a senior level of government," or "people who live in apartments," for example) will be stuck with the cost of correcting the situation. On the other hand if they were asked to contribute cash to some anti-pollution measure, the contribution to be equal to the damages they now suffer, many might underestimate their own damage, in the hope that others would contribute enough to put the plan into effect anyway.

Nor would any simple voting system serve to tell us whether the benefits of a proposed anti-pollution project would exceed its cost. Suppose the cost of the project is known to be $30, and there are three voters. Two voters decide that their damages are only $8 each, so that they are not interested in paying $10 (their share) to avoid them. The plan will be defeated at the polls, even if the third voter in fact suffers $20 damages, so that total damages of $36 could have been avoided at a cost of $30.

If individuals cannot accurately value the damage they suffer from different amounts of pollution, or are unwilling to reveal the amount if they do know it, is it possible for some outside person – some expert in "damage appraisal" – to supply the information we need? To some extent this may be possible, particularly in the case of private damage avoidance costs. Thus while I have only the haziest impression of what air pollution in Toronto costs me in terms of excess cleaning and painting bills, a statistician, by comparing per capita expenditures on these items in different cities and towns with different amounts of air pollution, could probably provide a better estimate than I could of these parts of my pollution damage. The estimates you see in the newspaper of "the cost of pollution" are often of this variety, and if carefully prepared they are useful so far as they go. They are not, however, complete estimates of damage avoidance costs. It is difficult to estimate expenditures made by firms to process the air in their plants so as to avoid damage that would otherwise be caused by air pollution, or to produce good figures for the damage done to automobiles by pollutants in the air; and it is all but impossible to find out how many families and firms have moved away from the city in order to avoid air and water pollution, and how much higher their transportation expenses are in their "second-best" locations.

Even greater difficulties of appraisal become evident if we try to measure the damage done by pollution to such intangibles as the recreational benefits and the direct aesthetic enjoyment of natural environments. Some economists have tried to produce a measure of the value of certain natural resources for recreation purposes. For example, one might try to get at least a minimum value of "Pike Lake" for recreation purposes by adding up the expenditures made by people who visit Pike Lake on fishing and duck-shooting licences, guide services, boat-rentals, fishing gear, ammunition, travel and living expenses, and probably several other things. Once you have done the addition, however, it is very difficult to know what the resulting figure means. Suppose there is another lake a hundred miles further away that is not at present used for recreation purposes. Is its value for recreation purposes zero? If Pike Lake became unusable, and all vacationers then travelled to the other lake, is it not true that the only value of Pike Lake to them was the savings they used to enjoy by not having to travel the extra 100 miles that they now travel? And if all lakes become unusable, and people turn from swimming to, say, folk-dancing, for their recreation, how can anyone say how much less enjoyment they get from folk-dancing than from swimming? (Those who put a very high valuation on swimming will, of course, offset their loss by building swimming pools rather than by taking up folk-dancing.)

At any rate, folk-dancing aside, it seems to me to be impossible to estimate the recreational value of an acre of water in a large area (Ontario) by calculating the per acre expenditures on water-based recreational activities at one particular location (Pike Lake) when there are many other locations in the larger area that offer similar facilities, and when the use made of each depends partly on the use made of others. We have chosen Ontario as our designated area for the study of pollution, and the choice of such a large area was made partly in order to stress the interrelatedness of pollution problems in different subenvironments. Because both men and wastes move around, it seems to me to be very important not to overlook this aspect of pollution problems. The sort of "recreation value" analysis we have been discussing, which is a part of benefit-cost analysis, is best suited to the study of a particular pollution problem in one small area, on the assumption that everything in every other part of the larger area remains unchanged. In that context the results may have some value, but if one attempts to include in benefit-cost analysis the cross-effects, as between different areas, of various pollution problems and anti-pollution projects the technique quickly becomes so complicated as to be virtually useless.

It may seem unfair to stress recreation because its value is admittedly hard to measure. But it is not unfair, because recreation, and such things as aesthetic enjoyment and freedom from health hazards, which are at least as difficult to value, are a very large component – probably the largest component – of pollution problems. Benefit-cost analysis was developed to deal with such problems as whether to build a dam to help control a river. The costs and benefits of such a project are *mostly* objective and can be either measured or estimated with some degree of accuracy. It is true that there are usually some subjective benefits associated with such projects, but they are normally such a small proportion of total benefits that it seems reasonable to give them relatively little weight in making the decision. Not so in pollution problems. Accordingly, benefit-cost analysis becomes much less useful; it becomes only an orderly way of setting down what we know and what we don't know; as a numerical aid to decision-making, it has little or no value.

AN ILLUSTRATION

Let us now try to illustrate some of the difficulties we have been discussing by imagining a realistic pollution problem. Let us say that there are complaints about the growing pollution of "Clear Lake." There are reports of people getting sick from drinking untreated lake water; swimming at several beaches is not as pleasant as it used to be because of an increase of algae in the late summer; and fishing is poorer than it used to be. There are plenty of "solutions" being proposed. We shall consider four of them: do nothing; build water treatment plants; build primary sewage treatment plants; build tertiary sewage treatment plants. What we want to know is which is the best policy; and we therefore start to draw up a benefit-cost table for the four proposals.

Can we calculate the value of the damage that is being done by the present polluting of the lake? We start out by trying to estimate the present level of damage avoidance costs. We know that some cottagers have already installed individual water-treatment devices; at considerable expense, we hire someone to make a survey to find out how many of these devices are in use, how many people plan to buy them in the next couple of years if the quality of the water in their neighbourhood continues to deteriorate, and the cost per year of such installations. Our man comes up with what seems like a pretty carefully calculated figure of $250,000 a year for the probable total amount of expenditure on this type of apparatus by two years from now. He found out, too, that some cottagers are boiling their drinking water at least part of the time, and that others haul their drinking supplies from farm wells, and even from the city. But several people he interviewed were evasive in answering his questions about these practices, and he felt unable to give reliable estimates of the expenditures people were making to boil or import their drinking water. He also reports that while making his survey he heard of several cottagers who had decided to spend their holidays elsewhere because they had become unhappy with the increasing pollution of Clear Lake. Newcomers are still buying the cottages of those who are leaving; but the sellers are finding that they can't get enough for their cottages to buy equivalent properties elsewhere. Our investigator said he would have liked to make an estimate of how much it was costing the defectors to avoid the pollution in Clear Lake, but he didn't think he could make a good estimate unless his budget were quadrupled – and even then he wasn't sure of success. Thus we decide to be content with an incomplete statement of present damage avoidance costs.

But for every person who is *doing* something about the pollution problem, there are six or seven who are complaining – and who are certainly suffering welfare damage – but who are not doing anything about it except to pay the odd medical bill when their children get sick (and they seem to be getting sick more often in recent years). And there are a few people, the new-comers especially, who don't seem to realize that they have a problem at all; many of them seem to come from the "Mud Lake" area, and think Clear Lake is the perfect answer to the pollution of Mud Lake. Thus we just don't know what the welfare damages are; all we know is that they are greater than zero.

When we turn to the question of remedies, the engineers tell us what the annual cost of the water treatment plant and the two sewage treatment plants will be, and also what their benefits, in physical terms, will be. The water treatment plant will provide for safe drinking water the year round (and will thus remove the health hazard from this source), but it will, of course, have no effect whatever on the water in the lake (and people may still get sick from swallowing lake water while they are swimming). The primary sewage treatment plant will remove solid wastes from

47

domestic and industrial sewage; but it will do nothing to remove dissolved materials, whether organic or inorganic. It will improve the quality of the lake water slightly for drinking and swimming purposes – or perhaps it would be more accurate to say that it would slow down the rate of deterioration in the quality of the lake water – but it would have no effect on the algae problem, which is becoming worse every year, and no appreciable effect on fishing. Tertiary treatment would remove a very high percentage of organic wastes; the lake water should be rendered safe for drinking and swimming. Tertiary treatment would also remove some 75 to 80 per cent of the inorganic salts, mainly phosphates and nitrates, from the sewage. It is these salts, deriving in part from body wastes, but especially from detergents, that act as fertilizers, or nutrients, for algae and have created the complex and obnoxious algae problem in the lake. Tertiary sewage treatment should significantly reduce this problem. It will not, however, reduce it by 75 or 80 per cent, the proportion of the nutrients removed from sewage, because an unknown portion – perhaps as much as a third – of the nutrients now entering the lake comes from farm run-off, and this source of supply would not be subject to treatment. (The nutrients in the run-off come from animal manure and artificial farm fertilizers.) Not enough is known about the mechanics of the nutrient-algae problem to permit any guarantee about the effect of tertiary sewage treatment on algal growth, but the engineers' best guess is that it would be reduced enough to make swimming much more pleasant, to improve fishing significantly, and to reduce greatly the stench of rotting algae along the shores of the lake. The estimated annual costs? $500,000 for the water treatment plant; $800,000 for primary sewage treatment; and a whopping $2,000,000 for tertiary treatment.

It is rather discouraging to find that even the engineering information, especially about prospective benefits, is so inexact. Of course, we reflect ruefully, even if the engineers could tell us that "the damage to swimming will be reduced by 57 per cent," rather than saying that "swimming should be considerably better," we would not be much better off. Since we have no way of estimating

48

Benefit-Cost Table III

1 POLICY	water treatment	primary sewage treatment	tertiary sewage treatment	do nothing
2 GROSS BENEFIT	drinking water only; more than $250,000 –probably much more	slight improvement to drinking water and swimming	drinking supplies safe; swimming much improved; fishing improved; algae problem significantly reduced	none
3 COST OF AVOIDING DAMAGE	$500,000	$800,000	$2,000,000	$0
4 NET BENEFIT (item 2 − 3)	unknown	unknown	unknown	unknown
5 WELFARE DAMAGE REMAINING	present danger to swimming, and health hazard of swimming; present damage to fishing; present algae problem	most present damage to swimming and health; present damage to fishing; present algae problems	part of present damage to fishing; part of present algae problem	all present damages

the present damage to swimming, we obviously cannot calculate 57 per cent of that damage. Indeed in most respects we are unable to attach dollar figures to either present damages or damages that would be avoided by the various proposals, and are therefore unable to state the expected benefits of any of the proposals in numerical terms. The only statistical data we have are the engineers' cost figures and our researcher's estimate of the costs people are paying for individual water-purification devices. We must, perforce, reconcile ourselves to a very "impressionistic" benefit-cost table. But let us draw up a table anyway and put down what we have learned about our cursed pollution problem.

The table helps us a little bit perhaps. It is certainly a systematic way of looking at things; we can see at a glance both what we get

(Row 2) and what we *don't* get (Row 5) for our money (Row 3). On the basis of the imaginary situation we have assumed for illustrative purposes, the primary sewage treatment plan seems very unattractive; it apparently offers very little at considerable expense, and most of us would rule out that project without further ado. But so far as choosing between the other two projects is concerned, or deciding to do nothing, we are not any farther ahead. If we only knew how much swimming and fishing and reducing the algal growth were worth and how much the people who don't buy water-purifiers would pay if they were relieved of the nuisance of boiling all the water they drink or the worry of drinking the water without boiling it! If only we knew these things, and all the other things one would like to know, the answer would be easy, as it was in chapter III. But we do not know the figures we need, and we do not know how to get them; so the table can't help us to choose.

Some people might say that they think the problem is greatly exaggerated – they neither swim nor fish; they only run their motor boats and sit in the sun – and claim that the total damages of pollution do not exceed $400,000 a year all told. Even the cheapest project, in their opinion, would cost more than it was worth. "Therefore," they say, "none of these projects is as good as the present policy of doing nothing." And no matter how much this contention angers us, we cannot prove it wrong. Moreover, those who favour some "active" policy disagree among themselves about which is the best project. But no one can demonstrate to others that his pet project really *is* superior to theirs. In the absence of hard, cold, agreed-upon numbers, opinions and subjective judgments hold sway. And everyone cherishes his pet opinions as dearly as everyone else cherishes *his*.

Would voting help us make up our minds? It is difficult to think of a fully satisfactory method of voting. At the very least it would be necessary to have a couple of elimination votes; otherwise some policy would probably win with a plurality but not a majority, and a majority of people would probably feel that they would have done better to compromise on some third policy which, even if not their favourite, would at least have been better than the one actually chosen. But even if we could think of a satisfactory method of voting, a most awkward problem arises: Who should vote? Permanent residents of the Clear Lake area, certainly; probably also summer residents. But how about those who might like to become either permanent or summer residents of the area if the lake were "cleaned up"? If they were willing to pay a fair share of the cost of cleaning it up, should they not have a say in the decision? Why not have everyone in Ontario vote on the issue, for everyone in Ontario surely has *some* sort of stake in the condition of Clear Lake? Many more awkward questions could be asked, but this tiring and negative sort of discussion can lead nowhere. Perhaps we should agree that voting is as useless as economics when it comes to finding the best solution to any real-world pollution problem.

POLITICAL SOLUTIONS

Politicians will of course find a solution to any problem that electorates insist they solve. Again, however, some politicians may conclude that nothing should be done and, if they are able to convince enough voters that they are right, they will be elected. Assuming that something *is* done, though, we may speculate a bit about what it might be, and what might happen as a result.

If Clear Lake's problems are deemed to be municipal problems, the people who vote on them will be municipal residents, probably both full-time residents and cottagers. If they choose the cheapest policy, the water treatment plant, everyone will be able to see that the condition of Clear Lake will continue to deteriorate and that cottagers who take a dim view of pollution will begin to move away to some less polluted vacation area. Those who do not object strongly to pollution will replace them; and Clear Lake will no doubt continue to satisfy their wants for many years to come. On the other hand, if the outcome of the municipal vote is that the most expensive policy, tertiary sewage treatment, is adopted, taxes will rise considerably and those who do not object to pollution as strongly as they object to taxes will begin to move out; they will be replaced by those who are willing to pay a stiff price to spend their vacations on a high-quality lake. Thus the outcome of leaving

51

pollution problems to be solved by municipalities is likely to be a tendency to the "natural zoning" of different areas; different cities and different vacation areas will probably differ significantly in their pollution levels (and in their tax assessments for pollution prevention schemes), and people will move around until they find the combination of water quality and tax level that suits them best. Average pollution levels over all municipalities might not change much, but after people had "paid their money and taken their choice" the volume of complaint about pollution would no doubt decline.

Until recently, both air and water pollution in Ontario have been under municipal jurisdiction; and the number of complaints probably has been kept down by people "sorting themselves out" into different areas. Certainly a lot of sorting out has gone on so far as vacation areas are concerned; and perhaps more people than we think have chosen their place of residence, and even their place of business, with an eye to pollution levels.

Such a solution to pollution problems, however, has limitations. As time goes on, and particularly if population and affluence increase, all areas are likely to become more and more polluted, and there will be fewer and fewer municipalities to which "pollution haters" can flee – at least at reasonable cost. Then an occasional magazine article appears, showing how the quality of one municipality's air and water depends not only on that municipality's expenditure on anti-pollution schemes, but also on what happens in one, or several, other municipalities – and the residents of other municipalities always seem to care less about pollution than *we* do! People begin to get a sense of claustrophobia, a feeling that, after all, sub-environments *are* linked together and that to some degree (and apparently to an ever-increasing degree) everyone does sink or swim together. Complaints about pollution – not only about where one lives or spends one's vacation, but about pollution everywhere – multiply enormously; and journalists and professors are not slow to cater to the demand! Very quickly, probably during some short interval of time when population grows so rapidly that space seems suddenly to shrink to uncomfortable proportions, a full-fledged pollution problem materializes.

But this time it isn't a municipal problem; it is at least a provincial problem and may well be a federal problem.

In Ontario certain aspects of water pollution became a matter of provincial concern in 1956 when the Ontario Water Resources Commission was established. On January 1, 1968, air pollution came under provincial jurisdiction. Meanwhile, the O W R C has expanded its activities, or at least its interests, to include most aspects of water pollution in the province; and it is now also concerning itself with certain aspects of soil and air pollution that seem likely to affect water pollution. In Ontario we have already graduated from "local" pollution problems, and are now well launched into the stage of "province-wide" pollution problems.

How is a senior government likely to respond to pollution problems? It is not, in my opinion, likely to encourage any "voting" solution to them. The reason why voting is not encouraged, I think, has to do with the awkward question that I raised with regard to our Clear Lake problem; it is simply too difficult to decide who should be allowed to vote for what. Senior governments are therefore more likely to attack pollution problems by administrative means. A department or a commission is set up and told to get to work – but not to do anything, of course, before the cabinet gives the green light. The responsible body – responsible to the cabinet, that is – has a horribly (or perhaps an entrancingly) difficult job to do. It cannot find the best solution to pollution, for as we have seen there *is* no best solution, even to any *particular* pollution problem. Yet it must decide what the problems are (no easy task in itself), arrange them in some sort of order of urgency (and it is always difficult to set priorities), and then start to do something about some of them (something that both makes sense to the engineers on the staff and is agreeable to the cabinet).

If some government body were required to solve our Clear Lake problem, I have a strong hunch it would choose the primary sewage treatment option, even though, according to our benefit-cost table, this seems to be the least attractive policy. The engineers would no doubt like to try the full-scale, tertiary sewage treatment; but the politicians are likely to argue that it is wiser to *start* with primary treatment and then see how things look, i.e., how the electorate seems to feel about both the lake and the increase in taxes. Perhaps in a few years the engineers will be allowed to proceed with full sewage treatment. In the meantime, with the primary treatment plant, the politicians will be able to claim that they are at least doing something to improve drinking water and swimming, and even (with some poetic licence) fishing; if they opted for the cheapest project, the water purification plant, they could not claim to be doing anything to improve the lake; besides, they might appear as cheap-skates! Again, therefore, on political grounds, the medium-price project, with problematical benefits, but a little something for almost everyone, seems like the best choice.

It is easy enough to condemn political decisions as inevitably inferior. But inferior relative to what? Our argument has led us to the conclusion that, in the pollution field at least, it is usually impossible to *prove* one policy inferior to another, no matter what our private convictions may be. Out-of-hand criticism of governments either for not doing enough or for doing too much about social problems is so easy precisely because social choice is so difficult. There is no best solution; to arrive at even a reasonably good collective decision about pollution problems will require not only an enlightened government but also an enlightened electorate; it will require also a free exchange of information and views among all parties concerned – governments, experts, community leaders, and the interested public.

By way of concluding this chapter, let me sketch out my own view of the steps involved in making a collective decision and taking collective action about pollution problems. Some of the matters raised will be dealt with more fully in later chapters of this essay.

The first stage in the process is reached when the volume of public demands that the government "do something" about pollution (i.e., do something different from what is now being done) becomes large enough that the government decides that the question should be "put on the agenda." Let us say that the government has been following a "do-nothing" policy. The first question it should now ask itself is whether the evils complained of result from a defect in institutions which could be corrected by a change in the law, or whether in order to deal with the problems effectively the government will itself have to become actively involved in setting up and administering pollution-control schemes.

This question, I suggest, is of the highest importance. Unless enforcement costs are expected to be very heavy, a change in the law is likely to be a far simpler and cheaper procedure than the establishment of a public pollution-control agency. Legal solutions to problems may not be very effective; the rise of the administrative state suggests, indeed, that legislative remedies are generally held in low esteem. Yet in such matters we are always in danger of being victimized by circular reasoning. Legal solutions to social problems may be weak and out of favour because little effort is being devoted to improving them – because they are out of favour! In any event, I raise the question because it is my impression that legal studies have for the most part failed to view the law as a social science, and that legal solutions for social problems are accordingly much less frequently proposed and discussed than they should be. The economist, who tends to view government as something like a business firm, and the public administration expert, who is interested primarily in the efficient working of organizations established to provide administrative solutions to social problems, deluge us with administrative schemes; it would make for wider choice and more competition in the market for ideas if the lawyers could tear themselves away from private law long enough to do some thinking about public law and put forward some legislative proposals for dealing with social problems.

Let us assume, however, that the outcome of this first stage of discussions is that effective control of pollution will necessarily involve direct administrative action by some agency of government. In the second stage of social action, the government is faced with at least three specific questions: What physical programs should be set in motion? Where should they be established?

How much money should be spent on them – counting both government expenditures and private expenditures that are required to be made as a result of action taken by the pollution control agency? Experts of various kinds will be able to prepare a list of projects that they recommend; each should be accompanied by a clear statement of expected costs and benefits, stated in numerical terms where possible. After careful consideration of the various proposals, the responsible authorities must arrange them into a list of priorities. That takes care of the first question for the moment. Some of the proposals will probably be province-wide; others will probably be proposals for a particular scheme in a particular area of the province; still others will be general proposals (for, say, sewage treatment) that could be applied in any area. The second question that the authorities must deal with is that of area priorities whenever a choice among areas comes up. These priorities will no doubt be arranged by the agency itself in terms of relative urgencies of the problems in different areas; but in the final analysis they will probably be juggled somewhat to take into account what governments are pleased to call "political realities."

The third question – how much should be spent on pollution control – will of course be decided primarily by politicians rather than by the agency authorities. This question is the heart of the matter; it poses the basic social choice to which, in my opinion, there is no scientific answer. I would suggest, though, that once the government has agreed that it should act on pollution it has a responsibility to act energetically. I think it should go for the *maximum* amount of expenditure that the public is willing to support. This amount it can find out only by a policy of escalation. Let it start with a modest program – the primary sewage treatment proposal in our Clear Lake illustration – and then increase it year by year until complaints about anti-pollution expenditure exceed complaints about pollution and make it politically dangerous for the government to go further. Only in this way, I think, can we approximate the cut-off point where people would prefer to bear the costs of pollution rather than the costs of doing anything more about it.

56

The third stage of public action should overlap the second and continue indefinitely. It should consist of a constant review of operations in order to find out whether the same results could be achieved more efficiently by alternative methods. Perhaps the lawyers will come up with legislative proposals that would achieve at less cost some of the results that are now attainable only at considerable expense in terms of both money and man-hours. Political scientists may make proposals about new methods of voting or new forms of regional government that have relevance for pollution control. And economists, given the present interest of the profession in the economics of "public goods," may also generate ideas that will help us to run our administrative state more efficiently.

I take for granted a flow of technical inventions and new ideas – automatic monitoring devices, new techniques for sewage treatment, and new industrial processes – that will facilitate social control of pollution problems. I have emphasized the possible contributions of the social sciences because I think they have much to offer in this area, even though in the past they have offered so little. While attention has been lavished on private business by economists, private law by lawyers, and traditional forms of government by political scientists, the administrative state has grown up under our noses. It has rendered obsolete the old debates about private enterprise *versus* public enterprise, laissez-faire *versus* planning, and capitalism *versus* socialism. This was the stuff of social science a generation ago, and social science has not yet fully adjusted to a world characterized by mixed economies and large governmental bureaucracies. But the old, over-simplified issues are dead and ought, in all decency, to be buried. It is high time that social scientists began to study the natural history of the administrative state and to suggest means by which we may keep it subject to human control.

The property interface

"Interface" is current academic slang for "boundary." It is along physical interfaces, when air meets water, or water meets land, or prairie meets woodland, that many of the most mysterious and exciting phenomena of the physical and biological world occur. Similarly in social science. The dividing line between work and leisure forms an interface between economics and sociology; behavioural studies at this margin throw considerable light on social attitudes on the one hand and economic performance on the other. And every observer of politics knows how important the "floating vote" is, how behaviour along the margins between voters and non-voters, and between party A supporters and party B supporters, affects the outcomes of elections.

"Property rights" form interfaces between law and several social sciences, especially economics, political science, and sociology. It is with property rights as the dividing line between law and economics that we shall be chiefly concerned, and our first task is to survey this boundary from both sides of the fence.

In everyday conversation we usually speak of "property" rather than "property rights," but the contraction is misleading if it tends to make us think of property as *things* rather than as *rights*, or of ownership as outright rather than circumscribed. The concepts of property and ownership are created by, defined by, and therefore limited by, a society's system of law. When you own a car, you own a set of legally defined rights to use the vehicle in certain ways and not in others; you may not use it as a personal weapon, for example, nor may you leave it unattended beside a fire hydrant. Among the most important rights you do have are the right to prevent others from using the vehicle, except with your permission and on your terms, and the right to divest yourself of your ownership rights in the vehicle by selling them to someone else. We may say, then, that ownership always consists of (1) a set of rights to use property in certain ways (and a set of negative rights or prohibitions, that prevent its use in other ways); (2) a right to prevent others from exercising those rights, or to set the terms on which others may exercise them; and (3) a right to sell your property rights.

What economics deals with is the buying and selling, or leasing, or using, of property rights. It could hardly be otherwise. You can only buy, sell, lease, rent, lend, or borrow things that are owned; and the only things that are owned are property rights. The prices of the things you buy and sell are prices for property rights to those things.

We can see immediately, then, the interaction that is constantly going on at the interface between law and economics. Consider further the example of automobile ownership. If property rights change so that automobiles cannot be driven unless they are equipped with exhaust-control devices the price of automobiles is likely to rise; if property rights are changed so that automobiles may not be used in cities, the price of cars is likely to fall. On the other hand, if the price of cars rises or falls, so that people own fewer or more of them, there will be social and political pressures on the legal system to change the prescription of property rights in automobiles – to change the law about where they may be driven, how fast they may be driven, where and when they may be parked, and so on. There is always, then, an interface where the law of automobile ownership and the economics of automobile ownership meet, but the boundary may be shifted by a change in either the legal or the economic aspect of the situation. So long as such shifts are small and gradual the interface remains relatively peaceful. But if a major change occurs in either the law or the economics of automobile ownership, the shift in the boundary may be large and abrupt; the legal and economic aspects of the situation tend to become disjointed, and the interface then becomes seriously disturbed. This may be as good a way as any of describing the emergence of a new social problem or set of social problems. But before we attempt to analyse such problems, we must pursue the concept of property rights somewhat further.

MORE PROPERTY RIGHTS

We have discussed property rights to physical objects. The concept can be easily extended to such things as money, stocks, and bonds. These documents have no particular uses in themselves, but they give their owners certain rights; these rights are exclusive – others can be prevented from using them – and they are transferable. In our society, too, individuals may be said to have property rights to their own labour: a person can use his time for any legal purpose; he has exclusive rights to the fruits, monetary or otherwise, of his efforts; and he can sell his services (and often use them as security for borrowing). In the "extended family" system that is found in several African societies people do *not* have full property rights to their labour since one's distant relatives have a customary right to share in one's earnings; the lack of exclusive property rights to income naturally results in some unusual features of the labour market in extended family societies.

It is when we come to various forms of public property that the concept of property rights becomes rather more tenuous – and yet, paradoxically, even more enlightening. A publicly owned building is very much like a privately owned building; the government can use it for whatever purpose it wishes, can prevent others from using it, and can sell or rent it. A public road system is a rather different matter; since it is built for public use (not, like a government building, solely for the use of government employees) there is no question of exclusive use, and in practice there is only very limited transferability of the asset. Thus while a government "owns" a road system, and can set general rules about its use, its ownership is clearly of a very special kind, reflecting the special public nature of the asset.

We also say that a government "owns" the air and water systems within its jurisdiction. Air and water create special problems partly because they are "natural" assets – unlike roads, they are not man-made and the quantity of them cannot be altered – and partly because they are mobile, "flowing" resources that move around from one area to another. About the only things in this world that are not owned in any sense are the high seas and their animal inhabitants. We shall call these special types of property – roads, water, air, public parks and so on – *common property*. The term covers all property that is both owned in common (or unowned as in the case of oceans) and used in common; and it is to a study of common property that I now turn.

Before I do so, however, I want to try to avoid a possible misconception. I have been talking about property rights and not "private property." That sadly overworked and ill-defined phrase, "private property," has become an ideological concept that I want nothing to do with. I think that in some cases property rights should be vested in individuals; in other cases in groups of individuals, such as firms; and in some cases in governments. Different types of situations, it is reasonable to suppose, call for different forms of ownership. But in any event: no ideology! No "private property"! Just "property rights," by whomever exercised.

COMMON PROPERTY, RESTRICTED AND UNRESTRICTED

The first question to ask is why some property is owned in common. If we think of the history of this continent we remember that at one time virtually all the land was owned by some government. The land, however, was sold off to private owners, except for some public domain that the government wished to keep for its own use, and for the land in far northern areas in Canada that no private party wished to buy. On the other hand, many road systems were at one time privately owned toll roads, and these have all been brought under common ownership. What makes it easy to arrange for full property rights to land, and thus for private ownership of land, is that land is easily divisible (you can buy a few square yards or a few square miles) and not mobile (by means of fences, exclusivity of use can be enforced at a reasonable cost). Roads are land, of course, and the existence of common property in road systems is therefore a matter of choice; people have made a collective decision that it is more convenient to build and operate roads in common than to have private owners run the "road industry."

Air and water (except for a few non-navigable streams), however, seem to be owned in common because there is no alternative. There is no feasible way of separating a cubic yard, or an acre, of water from other cubic yards or acres; there is therefore no way of ensuring exclusive use; fences simply don't work. (Major drainage basins provide for a physical separation of waters up to the point where they enter oceans. These large units might be bought privately by one owner; but even if a private buyer came forward, society would undoubtedly decide that it would be undesirable to put all the water in a major drainage basin under the control of a private monopolist.) Thus it seems to be the physical characteristics of air and water, the fact that they are fluids and are naturally mobile over the face of the earth, that make it inevitable that they be owned in common or "vested in the right of" some government as the constitutional lawyers put it.

The nominal owner of a common property asset (i.e., some government) has, of course, an undoubted right to lay down rules for the use of the property. Rules for the use of man-made common property, such as parks or roads, are usually promulgated by the owner; as is nearly always true of stationary property – the French, with reason, refer to real estate as *immeubles* – enforcement of rules about use is practicable at reasonable cost. Where specific rules about the use of common property are laid down, we can call it "restricted" common property. Until recently, however, most governments have *not* made rules about the use of air or water. Implicitly, the government policy or rule has been that anyone could use air and water for whatever purpose he wished,

62

without charge, permission, or hindrance. Such property we shall refer to as unrestricted common property.

In the past most people, at least in Canada, would no doubt have agreed that air and water resources were so vast that no rule about their use was needed. That day has now passed. But even if it be agreed that some regulation of use would be beneficial, we must still ask if the desired rules would be enforceable at reasonable cost. A "no-policy" policy makes sense if the cost of enforcing a positive policy is greater than its benefits. The costs of enforcing a policy fall, however, with improvement in administrative techniques, including such administrative hardware as computers and automatic monitoring devices. Such improvements therefore involve the possibility that rules about the use of common property resources that were impracticable in the past may be practicable now. Moreover, as larger and larger populations press against our fixed resources of air and water, the benefits to be gained by rules regulating their use increase, while enforcement costs are likely to become more easily bearable as they are spread over larger populations. These considerations suggest why our political scientists and our lawyers ought to be on a continuous look-out both for new legislative methods and for old ones that become newly practicable, in order to help control the use of our two most important common property resources, air and water.

The economic effect of making common property available for use on a no-rule basis, so that it may be freely used by anyone for any purpose at any time, is crystal clear. Common property will be over-used relative to both private property and to public property that *is* subject to charges for its use or to rules about its use; and if the unrestricted common property resource is depletable, over-use will in time lead to its depletion and therefore to the destruction of the property.

There is an old saying that "everyone's property is no one's property," the inference being that no one looks after it, that everyone over-uses it, and that the property therefore deteriorates. History bears out the truth of this saying in many sad ways. Property that is freely available to all is unowned except in a

purely formal, constitutional, sense, and lack of effective ownership is almost always the source of much mischief. The inefficiency
of medieval farming resulted in large part from the fact that
ownership rights were usually poorly defined; in particular, the
commons were unowned (i.e., owned by everybody – and nobody), and common pastures were so overstocked that their
productivity fell to the vanishing point. Not until ownership concepts had evolved to a point where something like a modern view
of property rights in land became accepted was it possible to use
the land efficiently and increase agricultural output.

Another example: The sad list of animal species that have been
extinguished by man's predation results purely from the fact that
property rights in these animals did not exist, perhaps because
they could not have been enforced if they had been established,
but in any event because they did not exist. If animals are sought
after they are valuable, and if they are owned those who seek
them will have to pay their owners for the right to kill or capture
them. Owners will charge a high enough price for the right to kill
their animals that some stock of animals will always remain; you
don't have to be an economist to know that it doesn't pay to kill
the goose that lays the golden egg. No domestic animal has ever
been threatened with extinction simply because domestic animals
are owned. Nobody owned the buffalo or the passenger pigeon;
and in recent years whales and kangaroos have been sadly victimized by the absence of ownership. If in the past Canadian
governments had said of trees, as they said of buffaloes and
passenger pigeons, that they belonged to everybody and everybody could cut them down free of charge, we may be sure that
there would be no lumber industry or pulp and paper industry in
Canada today.

With the rise of the automobile, the treating of road systems as
unrestricted common property has accentuated congestion problems and public toll-roads may be the best way of relieving them.
At any rate cities are beginning to learn that freeways seldom
make for free-flowing traffic, and that the building of freeways
soon increases traffic to the point where more freeways have to be
built. Medieval men who witnessed the overstocking of unre-

stricted common pastures would understand automobile congestion on unrestricted common roads. (Knowing that common
pasturing led to the deterioration of the livestock as well as of the
pasture, they might also observe with interest the deterioration of
the automobile stock resulting from, say, a hundred-car pile-up
on a California freeway.)

Air and water in this country, and in most other countries,
have been treated as unrestricted common property; so long as
they are so treated air and water pollution will increase and the
physical condition of our air and water assets will continue to
deteriorate. Moreover, as has already been pointed out, we can
manufacture more roads, but we cannot manufacture more air
or water; all we can do is to use existing supplies as wisely as
possible. It is time, I believe, that we took air and water out of
the category of unrestricted common property, and began to
establish some specific rules about their use or, to put it another
way, to establish something more sophisticated in the way of
property rights to their use than the rule that "anything goes."
That rule may have been quite sensible in the past, when the
demands made by human populations on the services of air and
water were very small compared to the volumes of these assets;
the benefits of controlling use would probably have been small
and the costs of enforcing restrictions would no doubt have been
large. All I am arguing is that growth in population, production,
and urbanization inexorably changes the balance of the benefit-
cost analysis against the policy of doing nothing and in favour of
some positive policy. But to say that is not to say or in any way to
imply that it is an easy matter to choose a *wise* positive policy, or
to establish wise new property rights to the use of air and water.

SOCIAL PROBLEMS

Like benefit-cost analysis, an analysis based on property rights
provides a way of looking at pollution problems (and other social
problems); but unlike the economic analysis, which is confined to
the study of solutions that have been proposed, the legal analysis
often generates proposals for solutions to social problems. In this
section, I propose to look at, and comment on, a few examples

of the relationships between property rights and social problems. Two points should be kept in mind throughout. First, there is no perfect legal solution to social problems, any more than there is a perfect economic solution. Second, a given legal definition of property rights in an asset has not only economic consequences (as we have seen in the previous section of this chapter) but also social and political consequences; there are interfaces between law and sociology and law and political science, as well as between law and economics.

Consider, first, a frivolous example. You are driving in a city after a heavy rain, and inadvertently drive through a large puddle of water so fast that you thoroughly drench some unfortunate pedestrian who happened to be in the wrong place at the wrong time. When you notice through your rear-view mirror that the victim is taking down your licence number, you stop and, after a brief conversation, pay him $10, shake hands, and go on your way.

This problem, then, was settled expeditiously and with a minimum of social friction. The reason is that the legal situation was clear and known to both parties. Ownership of an automobile did not confer the right to dirty other peoples' clothing, and ownership of clothing did confer the right not to have it dirtied by inattentive motorists; and neither party was interested in having a judge tell them what they both knew. The law might, of course, have said the reverse – that pedestrians had to look out for splashing motorists, rather than saying that motorists had an obligation to avoid splashing pedestrians. Had this been the case, there would have been no more social friction, but the economic outcome would have been different; the cost of the incident would have been borne by the pedestrian rather than the motorist. If the law had not been clear, there would have been bad feelings, there might have been a court case, and the economic outcome would have been unpredictable. There is much to be said for definiteness especially where the law is concerned.

Notice, too, that the existing law about splashing problems imposes the cost of injury on the active party, the motorist, rather than the passive pedestrian. The probable rationale for this policy is not the *social* view that the pedestrian is more important than the motorist, but the *technological* consideration that motorists can more easily avoid splashing pedestrians than pedestrians can avoid being splashed by motorists, and the *economic* consideration that it is cheaper to persuade motorists that it does not pay to splash pedestrians than to protect pedestrians by building a six-foot wall along the interface between sidewalks and roads. As a pedestrian, a motorist, and a taxpayer, I think the present law is very sensible.

Imagine, now, that you own a factory in Toronto and that you have been dumping your untreated factory wastes into Lake Ontario for forty years. Until recently not a single person complained of your practices and you are breaking no law by continuing to do what you have always done. Yet in the last couple of years it seems that you have become a villainous polluter, a heartless despoiler of nature, and a sneak thief robbing the children of Toronto of their natural right to swim in Lake Ontario; the press is after your scalp and trying to put the government on you; even your best friends seem to think that "something ought to be done about pollution." You object to such rough treatment, and reply that your lawyers advise that you have as much right to dump your garbage in Lake Ontario as any kid has to swim in it. You are probably right, legally, but you are in for a lot of trouble with your public relations.

Two comments suffice. Unrestricted common property rights are bound to lead to all sorts of social, political, and economic friction, especially as population pressure increases, because, in the nature of the case, individuals have no legal rights with respect to the property when its government owner follows a policy of "anything goes." Notice, too, that such a policy, though apparently neutral as between conflicting interests, in fact always favours one party against the other. Technologically, swimmers cannot harm the polluters, but the polluters can harm the swimmers; when property rights are undefined those who wish to use the property in ways that deteriorate it will inevitably triumph every time over those who wish to use it in ways that do not deteriorate it. Economically and socially the question is always which set of interests *should* prevail, or rather what sort of accommodation should be made among the various interests concerned. The question is always, and inescapably, the great question of social justice.

Questions of social justice can be answered in many ways. Consider carefully the following example of an actual solution to water pollution problems in Britain. For this example, which I find utterly fascinating, I am indebted to Douglas Clarke who has recently described it in the following words.

The island of Great Britain is moist and verdant, and blessed with innumerable cool streams that once were all haunts of trout and salmon. Most of them still are, even though they now flow through an industrialized countryside. The total poundage of fine game fish taken would put any accessible part of Canada to shame. We are so used to the idea that the waters of any industrial area are a write-off, so far as quality angling is concerned, that one cannot help but be curious as to how all that fishing is maintained.

It is not because they do not have to watch out for pollution. There is an organization called the Anglers' Cooperative Association which has been in existence for nineteen years, which has taken over the watch dog functions formerly left to individuals. It is an interesting organization. It has a fluctuating and rather small list of members and subscribers, barely enough to keep an office open, but it is able to call on some powerful help, especially legal. It has investigated nearly 700 pollution cases since it started and very rarely does it fail to get abatement or damages, as the case requires. These anglers have behind them a simple fact. Every fishery in Britain, except for those in public reservoirs, belongs to some private owner. Many of them have changed hands at high prices and action is always entered on behalf of somebody who has suffered real damage. It has been that way from ancient times. Over here the fishing belongs to everybody – and thus to nobody. The A.C.A. exists merely to take action where individuals may not act themselves.

Two cases from some time back well explain why the Derwent, which flows through the industrial city of Derby, still has its trout. Action was entered against the city because its effluent was harmful to trout, and the city, through its legal representatives, claimed in the highest court in the

68

land that it was completely unreasonable to expect them to maintain the standards of a trout stream. The A.C.A., incidentally, acted on behalf of the "Pride of Derby Angling Club," which leased the fishery from the titled gentleman who owned it. The law lords said that the city had no more right to put its muck in the river than the citizens had to put theirs on the property of their neighbours. About the same time, and for the same city and river, an injunction was obtained against British Electric, a public corporation. All they had been doing was to run warm water directly into the river. Trout like it cool. The A.C.A. also deals with such – to us – trivia as mud running into a stream from a new road grade, or a ditch. It doesn't have to and the anglers are willing to go to court. This is actually a good example of a common form of pollution which we accept but which is quite unnecessary and not hard to avoid.

What it amounts to is that you can have good fishing, which means good water, in a river in a populated British countryside if you make it your business to have it. It is not only Britain. We get an anglers' magazine from Germany and there are lovely illustrations showing good fishing on the Ruhr river, of which you may have heard, and on the Binnen, or inner, Alster, ... in the industrial part of Hamburg. ...

I will be the first to admit that there are geological and climatic differences between Ontario and western Europe which have influenced the impact of European settlement on our area, so that some of our streams have, inevitably, a less constant flow and a warmer temperature than they used to have. Within these limitations, however, we ask ourselves why we have to sacrifice water quality still further by deliberate pollution.

Some time ago the A.C.A. analyzed their comparatively few failures. In some cases the polluter could not be identified. In some other cases the polluter was insolvent, hence no damages. They call this failure. However, and this underlines the comparison between them and us, the most important single cause of failure was when the anglers who suffered from the pollution had no concrete evidence of interest, such as a valid lease, and had only tacit consent or a gentleman's agreement with the owner, who refused to become involved in the action. That sounds familiar. We, as individuals, fish the waters that we all own, collectively. As individuals we have sustained no damages at law. Collectively – as owners – well, forget it. In Britain, when a truck involved in an accident spills chemicals into a stream, the public liability insurance pays for the fish, for all the costs of clean-up and restocking, and for the loss of use and enjoyment during the period between kill and restoration because property damage has been done. Who looks after us?

Officially we have tried to do by statute what the British have done by the Common Law, but never, apparently, have we really meant what we said. Our first legislation, in 1865, had its teeth pulled in 1868. It is interesting that one simply cannot conceive of a judgment or an injunction obtained through legal action by the A.C.A. being set aside. Part of the explanation may be social. The A.C.A. has the Duke of Edinburgh for Patron. Apparently it is quite all right for him to be honorary keeper of a watch dog that has sunk its teeth into government corporations such as British Electric and the Coal Board, municipalities big and small, industries and private individuals, without fear or favour. I notice that His Grace the Duke of Devonshire is President, and there are two more dukes among the vice-presidents, (that is over ten per cent of the total number of non-royal dukes), as well as two additional peers, [and] a couple of knights. ...

There are many worthwhile comments to be made about this passage, but let me mention only a few of them. Note, first, that the solution results from a particular set of property rights (based in this case on Common Law) that are enforced by the courts, and that the property rights seem to be in the fish, or the fishing, not the water; there is no administrative agency that is concerned either with the fish or the water. Second, the solution may seem simple but in fact it isn't; Mr. Clarke is careful to suggest that the workability of the system may depend in important ways on such apparently irrelevant factors as the English climate, English history, and the particular social status enjoyed by the English nobility.

Third, there is no way of knowing whether the solution is a "good" one or not. At one level of analysis it can be argued that the solution favours the fishermen over industrialists and municipalities who have to bear the costs either of disposing of their wastes in such a way as to avoid polluting the rivers or of buying up the fishing rights to a river and then using it for waste disposal purposes; and there seems no obvious reason why the shoe should

70

not be on the other foot – why polluters should not own property rights in the waste disposal capability of the river, in which case the fishermen, if they wanted to fish, would have to buy out the polluters' rights. Note, however, that if the government "owner" of the river follows a policy of "anything goes" neither party can buy out the other because neither has anything to sell! Under a system of unrestricted common property, groups that have opposing interests in the use of the property cannot negotiate because they have nothing to negotiate with; all they can do is yell interminably at each other.

In my opinion, however, it is often misleading to think of pollution problems in terms of groups rather than in terms of the society as a whole. In the present example fishermen no doubt live in cities and buy manufactured products, and industrialists and residents of cities no doubt sometimes go fishing. The groups are, in fact, all mixed up together. And it is not true that fishermen pay nothing for their good fishing; they pay higher prices for manufactured goods and higher municipal taxes than they would pay if the law favoured polluters and if fishing were not so good. Similarly, polluters get better fishing for their higher expenditures on waste disposal. From an over-all, social point of view the whole British population in effect buys good quality fishing (and other water-based recreation) by paying higher taxes and higher prices for goods; in Ontario, we have in effect accepted water pollution in return for cheaper goods and lower taxes. Which is the better policy? A silly question deserves a silly answer: whichever policy is preferred is better.

Mr. Clarke makes it quite clear that he prefers the British solution, or something like it. So do I. It would help if we could let our provincial member of parliament know roughly where each of us stands on the question of better quality air and water versus higher taxes and higher costs of goods. But it wouldn't help very much. The important question is *how much* "better quality environment" we would be willing to buy at different "prices" in terms of higher taxes and higher costs of goods, and most of us are not sure about this. As was suggested in the last chapter, the

71

only way to answer the question may be to have the politicians start charging us for better quality air and water and then keep "upping the ante" until we say "Enough! No more!"

The trouble is that when we call a halt about half of us will think we are already spending too much to improve the environment, and about half of us will want to spend more; therefore very few of us will be very happy with the outcome. In some cases there is nothing more that can be done. In many other cases, however, there *is* a better solution; we have in fact often adopted it, but it is only recently that Professor Mishan, an economist, has generalized the argument that underlies it. The point is that it is often possible to avoid the sort of fifty-fifty compromise that we have been discussing. Take the question of smoking on a train, for example. If all passengers, half of whom are smokers, are required to come to a single decision, they may decide to allow smoking or not to allow smoking (in which case half of them are going to be unhappy all of the time) or they may decide to allow smoking during half the journey (in which case all of them will be unhappy half of the time). The sensible solution in this case is to provide what Mishan calls "separate facilities," i.e., to provide both smoking cars and no-smoking cars; everyone should then be happy all of the time. This solution, of course, is not applicable in a single-cabin vehicle such as a bus; "separate facilities" may not always be practicable. Zoning laws in cities are another common example of the "separate facilities" type of solution.

Is this solution applicable to pollution problems? Not perfectly, certainly, but to some considerable extent. Although air and water move around they do not mix thoroughly; over a large area such as Ontario it is certainly practicable to provide for different air and water qualities in different regions. We noted in chapter IV that when pollution matters come under municipal control different municipalities, or groups of municipalities, are likely to provide different quality environments. Under provincial control, the same variety is possible if the provincial authorities choose to follow some variation of the "separate facilities" or "zoning" principle. This principle will not always be applicable, and may not always be desirable. It is, however, always worth considering, if only because it offers some possibility of meeting a variety of demands and opinions with a variety of solutions. Instead of giving property rights in water use to polluters *or* fishermen, it may be thought desirable to assign the rights to fishermen in one area and to polluters in another.

SOIL POLLUTION AND INAPPROPRIATE PROPERTY RIGHTS

As new products, new practices, and new situations emerge, property rights need continual redefinition. Torontonians with trees on their lots used commonly to burn their leaves every fall. As general air pollution increased, it was considered unwise to add to the burden of the air by the widespread burning of leaves. Accordingly Torontonians agreed to reduce their property rights in their leaves; they can now do almost anything they like with them, except burn them; normally they pay the city, through their taxes, to haul them away.

New technology, especially in the form of chemical fertilizers and pesticides, has posed certain serious problems of soil pollution. Too much fertilizer can render soil infertile, and unless it is carefully applied large quantities of it run off the land and add to the "excess nutrient" type of water pollution. The new pesticides, based on chlorinated hydrocarbons, raise new problems; because they are chemically stable, they accumulate through time. It is feared, though so far as I know not firmly established, that in sufficient concentrations they might kill enough micro-organisms to disrupt normal bacteriological processes and seriously reduce soil fertility.

It is certainly not the absence of property rights that is responsible for the threat of soil pollution, since full property rights exist both in the land and in the chemicals. Nevertheless property rights are not static things, and we may at least describe the soil pollution problem in terms of *inappropriate* property rights in the new chemical substances. When first introduced in the late 1940s, these toxic materials were considered to raise no particular question of property rights. They *might*, of course, have been considered potentially dangerous products – like guns, narcotics, some

73

medicines, and radioactive isotopes, for example – and it would then have been appropriate to define explicitly the property rights attached to them, i.e., what an owner could and could not do with them. (Remember that ideological overtones of such concepts as ownership or property are completely irrelevant to the present argument; ideological commitment to private ownership, though fairly strong in North American society, has not constituted much of a barrier to a fairly prompt, and restrictive, definition of property rights in hallucinatory drugs.)

It was almost certainly as a result of ignorance that when chlorinated hydrocarbons first appeared on the market their owners were given full property rights in them. Evidence of their unwanted effects on the environment built up, however, and after the publication of Rachel Carson's *Silent Spring*, which gave the public a few lessons in elementary ecology, the tide turned, and a general movement toward a narrower definition of property rights in these materials has been apparent. In the meantime, an impressive array of vested interests had quickly built up around both the manufacture and use of these products. It is hardly too much to say that the whole structure of agriculture had changed to take advantage of the great increases in productivity that their use made possible, and it seemed unthinkable to do anything that might jeopardize such recently won progress. The difficulty of *reducing* property rights in the interests of combatting soil pollution seemed as great as the difficulty of *establishing* property rights in the interests of combatting air and water pollution.

Nevertheless, in Ontario and elsewhere, we are now witnessing early attempts to tackle both problems. Not much has been done in Ontario about reducing property rights in pesticides; but publicity about their harmful effects and increasing awareness of their dangers on the part of government officials have already resulted in greater caution in their use. The pesticides case is indeed an interesting one from the standpoint of public administration. Except perhaps for radioactivity, which was recognized to be deadly from the start, mankind had never before confronted himself with man-made wastes that were both toxic and persistent.

74

Chlorinated hydrocarbons were something new in history, and governments had never before been faced with the types of problems they raised. It is not surprising that we made mistakes. I think it likely, though, that we have learned something by those mistakes. At least we can hope that as new man-made chemical wonders become available we and our governments will give long thought to the question of what property rights their owners should be granted. In the other direction, that of *establishing* property rights in air and water, much more has been done. For some years the province of Ontario has been pre-eminent in tackling water pollution problems; and in January 1968 provincial authorities took over jurisdiction from the municipalities in matters of air pollution, no doubt as a prelude to a more active policy in this field.

SUMMARY

This chapter has tried to suggest that legal definitions of property rights lie at the heart of social decision-making and problem-solving. Property rights are clearly antecedent to economics, since it is property rights that define the economist's "goods and services," and we have seen, particularly in the discussion of unrestricted common property, how property rights affect individual and social behaviour. A study of property rights gives us no magic key for the solution of social problems, but it does lead to suggestions for solutions that are refreshingly different from those offered by economists and other social scientists.

The main substantive conclusion I wish to draw from this chapter is that to treat air and water as unrestricted common property is socially indefensible. A policy of "anything goes" is defensible if the cost of enforcing a positive policy exceeds the benefits to be gained from it. This may be true now of polar ice, and it may have been true in the past of air and water, although I doubt it; English courts have apparently long enforced property rights in fishing, and in the process have enforced one solution to the problem of water pollution. In any event, it is perfectly clear on both theoretical and historical grounds that, as population

75

grows, unrestricted common property will be over-used and deteriorate physically to the point of uselessness. On the assumption that we don't want that to happen to our air and water, it is high time that we began to devise some new forms of property rights, not to air and water, but to the *use* of air and water. In Ontario, during the last dozen years, we have begun to move in that direction, at least to the extent of changing the status of air and water from unrestricted to restricted common property. But the field for new ideas and social experimentation is still wide open.

In the next chapter, I seek to show what economics has to contribute to the question of managing our fixed supplies of air and water. Since economics is based on property rights, economic solutions to pollution problems also involve property rights solutions.

Pollution rights

In the last three chapters I have argued that whether we approach pollution problems from the standpoint of economics (benefit-cost analysis) or the standpoint of law (property rights) we can find no best solution to them; and that any anti-pollution policy is therefore bound to be in the nature of a social experiment that is neither right nor wrong, but only more or less successful in leading to wise and socially agreed-upon patterns of use of our air and water resources. In this chapter I shall begin by sketching out one possible policy (or social experiment), and then discuss, in much greater detail, its implementation. Throughout this chapter I shall refer only to water-pollution problems; but the argument should apply equally well to air-pollution problems.

A POLICY

I would have the Ontario government set up a Water Control Board, which I shall call the W C B for short. The legislative function of the W C B would be to decide what the quality of all natural waters in Ontario should be; its executive function would be to implement its decisions. Both functions would, of course, be subject to over-riding review and veto by the government of the day; but the intent of the legislation establishing the W C B, like the intent of legislation establishing all such boards, would be to insulate it from party politics. Being largely ignorant of the principles of public administration, I shall sidestep the important questions of how many Board members there should be, how long each should serve, and what their qualifications should be. (On the latter question, though, I don't think the members should all be experts or all amateurs; a mixture would be more to my liking. *None* of them should be a party hack.)

At its first meeting, the Board members will unanimously agree that their terms of reference, "to decide what the quality of all natural waters in Ontario should be," are quite absurd if interpreted literally. I would advise them to divide Ontario into "regions" and then try to set an "average" water quality for each region. Suppose they adopt this strategy. After long discussions about whether they should define regions as "watersheds" or "population clusters" or "groups of municipalities," or some combination of these and other criteria, they *do*, however, succeed in drawing a map that divides the province into "water control regions." Their major policy task is now to decide what the average water quality in each region should be.

They run into one dismaying difficulty right way; there is no measure of water quality! Scientists talk about water quality in terms of "suspended solids," "B O D," "dissolved oxygen," "nutrient content," "pesticide residues," and half a dozen other things. But there is no over-all measure of quality; it turns out that even "pure" unpolluted waters differ greatly in terms of most of these measures so that it is impossible to define "pure" water and therefore impossible to define "impure" water or to measure "the degree of pollution."

Well, there is no use arguing with science. But there is no use, either, in arguing with several millions of people who insist that there *is* pollution and that the W C B do something about it. The Board therefore tries a new line of attack. "Suppose," some members say, "we look at the problem the other way around. If we can't measure water quality we can at least measure the amount of waste that is dumped into natural waters; and that, after all, is what people mean by pollution." The experts agree that, at least in principle, the number of tons of waste that are put into the water *can* indeed be measured. They raise two sorts of difficulties, however, that the Board must consider before they try to set their policy in terms of "tons of waste."

The first is that people are not interested in tons of waste but in the damage done by wastes; a ton of pulp liquor does different *sorts* of damage, and probably a different *amount* of damage, than a ton of untreated sewage. To get around this problem, the Board agrees that it will have to draw up a "table of equivalents" for different types of wastes, so that they may be expressed in some common denominator as it were. But it is immediately pointed out that a ton of any particular kind of waste will do much more damage in some places than in others; the damage done will depend on the particular character of the water, since some rivers and lakes have a greater capacity for absorbing wastes than others, and on the particular uses people make, or would like to make, of the water in question. A tentative way around this difficulty seems to be for the W C B, in consultation with its engineers, to draw up a *different* table of equivalents for each region, so that regional differences both in water and in water use can be allowed for, at least in a very rough and ready way. This solution will also enable the Board to make some allowance for the phenomena of counter-acting and escalating combinations of waste. Wastes that counter-act one another in one area may be given low damage ratings, though each is given a higher rating in areas where the other is not present; the reverse allowance may be made for escalating wastes. So far, the procedure seems practicable, if not very precise.

The second difficulty, the experts report, is that they can measure without too much difficulty the amounts of waste at identifiable "waste outfalls" – where pipes enter the water – but they really cannot measure with any accuracy the amount of wastes that enter water systems from surface run-off or from fall-out from the air. However, the experts agree that, in water pollution at least, most wastes come from identifiable sources, and the Board therefore decides to defer the run-off problem, and to deal initially only with pollution from waste outfalls.

The W C B can now, I think, feel that it has made some progress toward enunciating a policy; or, rather, it has found a way in which its policy, when formed, can be enunciated. True, much

79

remains to be done. A table of equivalents has to be drawn up for each water control region; this work will have to go forward in consultation with scientists who know their wastes and their waters, and with local interests who know the uses that people make of the water in their areas. There then remains the crux of the matter: How many "equivalent tons" of waste will the Board allow to be dumped in the water in each region? That is where the crunch comes. We have already seen that there is no "correct" answer to the problem; there is certainly going to be no unanimity to whatever answer is given; but the Board's main function is, precisely, to answer it.

Suppose the Board, after long deliberation, finally makes the following announcement: "The w c b has decided on a water quality control policy for a trial period of five years. During each of the next five years the equivalent tons of waste injected into the water in each water control region must not exceed the equivalent tonnage of wastes that were injected into the water last year. The Board will announce shortly how it intends to implement this policy. During the five-year trial period the policy will be subjected to continuous study and appraisal, and at the end of five years a new policy announcement will be made."

And so we have a brand new water-pollution policy! But how will it work? Vacationers may wonder whether the policy will permit new cottages to be built at the lake during the next five years. New cottages could be built if they were equipped with septic tanks, or if present cottagers built a small sewage treatment plant, reduced the present inflow of wastes entering the lake, and thus "made room" for newcomers. The same would be true for industry or urban housing; by treating present wastes more fully, growth in population and production could be allowed for without exceeding the present inflow of wastes to natural water systems. To that extent, at least, the policy sounds practicable. It will not, of course, do anything to *reduce* pollution (though the policy sets a maximum waste discharge, and there is nothing to prevent people from discharging less than this maximum) but it should help stop the *growth* of pollution (except that part of the problem

80

that derives from run-off and fall-out). The policy therefore seems to be helpful – if it works. It is to the interesting question of implementing their policy that the w c b, and we, now turn.

THREE POSSIBLE WAYS OF IMPLEMENTING A POLICY

To an economist, there are only three basically different ways of implementing the Board's policy. The first we shall call "regulation." The w c b can issue a regulation requiring all factories and municipalities to reduce their discharges by, say, 5 per cent (to allow for growth) or it can set an allowable quota of waste, expressed in equivalent tons, to each outfall, and simply decree that that quota shall not be exceeded. In the latter case, of course, it must make sure that the sum of the individual quotas is no greater than the over-all figure it has established for the region.

The second technique can be called "subsidization." We suppose that the w c b has direct access to "unlimited" provincial funds, raised either by taxation or by borrowing. The Board could, then, decide to subsidize whatever expenditures were necessary to keep the amount of wastes down to the figure it has chosen for each region. It could subsidize municipalities by building sewage treatment plants for their use; where factories discharge their wastes directly into water systems (rather than indirectly through municipal sewers) the Board could pay the cost of linking them up to municipal sewers or of installing waste treatment systems in the factory. The Board would then have direct control over the measures (and expenditures) required to ensure that its own policy objective was achieved.

The third technique we shall call "pollution charges." The Board, under this system, announces that it is going to levy a "disposal fee" on all those who dispose of their wastes into natural water systems. The fee per ton of waste may vary at different outfalls, and the charge may also increase as the number of tons of discharge increases. This technique is based on the principle that if you charge a person for disposing of his wastes he will find ways to reduce the amount of wastes he disposes of, and that the

81

more you charge him the stronger the incentive he will have to find some less damaging method of disposing of his wastes. The Board, however, may have to do quite a bit of "trial-and-error pricing" before it hits on a system of charges that results in the total amount of wastes that are discharged into water systems being equal to, or slightly less than, its target figure.

There are many fascinating and important comparisons and contrasts to be made between these three broad methods of enforcing our water-pollution policy. What is especially interesting is that both the techniques themselves and the analysis of them that follows apply to *most* social problems and are by no means confined to pollution problems. We cannot here discuss other social problems; but perhaps the reader might entertain himself by applying this kind of analysis to, say, aircraft safety standards, traffic problems, or building codes. It is interesting to note that the parking meter, as a solution to parking problems, exemplifies our third technique: a fee is levied for the right to park on public property, and different fees are levied on different streets. Traffic problems are controlled mainly by regulation and partly by charges (fines), and many educational problems are dealt with by subsidies. But, in principle at least, the technique that is actually used in each case could be replaced by either of the other two.

Note, first, that even though we have referred to our three techniques as "basically different" they in fact appear to amount to much the same thing. The w c b's policy can be met by a particular set of regulations, a particular set of subsidies, or a particular set of charges. "But who pays?" you will ask. Actually, we answered that question in the last chapter, when we were discussing the cost of the English system of "property rights in fishing" to the fishermen and the municipal or industrial polluters. If people are divided into groups, different control techniques have different results; under a subsidy scheme the polluter receives money (or equipment) while under a changing scheme he pays out money. However, if we carry the analysis a step further, we see that in a charging scheme producers and municipalities will recoup the money they pay out by charging higher prices and higher taxes for their goods and services; the same will be true of the higher costs industries and municipalities will be forced to bear if the W C B uses regulations rather than charges. If subsidies are used, prices of goods and municipal services will not go up, but provincial taxes will. In the end, the costs will be spread around, and the general population will pay for pollution control. This is why, when we are dealing with a large population and a large area such as Ontario, it seems more realistic to deal with society as a whole, rather than with groups. If we were dealing with, say, a pollution problem caused by one factory in a dormitory suburb, it would make sense to distinguish the factory group from the residents, and such problems are of course important. Here, however, we are dealing with what economists call "general equilibrium" situations, in which we are all simultaneously producers *and* consumers, polluters *and* pollutees. It is then true that, no matter who passes the money to whom in the first place, we all pay in the end. (Whether we pay *equally* depends on a host of factors, such as the taxation system, individual consumption habits, and so on; but there will be individual discrepancies in the burden of pollution control no matter how the control is implemented.)

Astonishingly, therefore, whether you pay industries (and people) not to pollute, or charge them for the right to pollute, or simply tell them that they must pollute only so much and no more, the outcome is still much the same. Pollution can be kept to the same amount by any one of the techniques, and everyone pays for keeping it to that amount. It is largely a waste of time for the pot to call the kettle black where pollution problems are concerned; everyone pollutes and everyone pays for not polluting.

Does it, then, make no difference which technique the W C B uses to enforce its policy? It certainly *does* make a difference, an enormous difference; even though the costs of getting the benefit are shared, the amount of cost (and also the type of cost) differs greatly from one scheme to another. Consider a simpler case for a moment. Suppose that we wanted to decrease the number of high-school drop-outs by, say, 90 per cent. We could do so by paying prospective drop-outs whatever price would lead nine out of ten of them *not* to drop out; or by charging them a fee for

83

the right to drop out that would result in only one out of ten prospects deciding to do so; or by passing a law forbidding drop-out (but allowing kick-outs). The paying scheme would be very costly to the taxpayer in terms of money because every student worth his salt would think of dropping out in order to collect his payment for deciding not to! The law forbidding drop-outs would, I think, be worse; those who wanted to drop out badly enough would have to do enough damage or otherwise make enough of a nuisance of themselves to be kicked out. The taxpayer would have to pay for the damage, or for controlling the nuisance. Much more serious, though, would be the feelings of resentment at being forced to stay in school by those who would otherwise have dropped out, and even by those who would like to think they *could* drop out if they wanted to; the cost in terms of educational morale might be high. The *charging* scheme would cost practically nothing to administer; those who still decided to drop out would actually be a small source of revenue to taxpayers; and there would probably be a lot less idle talk among students about dropping out than there is now – those who wanted to drop out badly enough would simply pay their exit fee and leave. A rather similar analysis applies to our three techniques of pollution control, though here there are many more things to consider.

TWO VARIANTS OF THE BASIC TECHNIQUES

Whether regulations, subsidies, or charges are used to implement a pollution policy, there is a choice between what I shall call "across-the-board" schemes and "point-by-point" schemes. By the former I mean regulations, or subsidies, or charges, that are applied uniformly to all polluters or all sewage outfalls; by the latter I mean schemes in which regulations (subsidies, charges) are adjusted to suit the circumstances of each individual polluter or outfall. In practice there are some across-the-board schemes that blend into point-by-point schemes, so that there is really a whole spectrum of ways to implement a policy. Our classification of all possible schemes into six types is designed to illuminate the main features of a variety of techniques actually used by pollution control agencies. Throughout, we shall suppose that the immediate goal of the W C B is to reduce the number of equivalent tons of waste currently being discharged into water systems by 5 per cent.

Across-the-board regulation would be most unfair. It would also be inefficient in the economic sense, i.e., the same result could be achieved at lower cost by some other scheme. Suppose that the W C B issues a regulation saying simply that "All existing factories and municipalities must reduce the amount of waste they discharge into water systems by 5 per cent next year." Those factories and municipalities which were *already* spending a good deal of money to treat their wastes would be justifiably annoyed; they would argue that those who now did *nothing* about their wastes should be the ones to carry the main burden of the over-all reduction in waste discharge, and that only when everyone is in the same position should everyone be treated in the same way. Moreover, municipalities on large fast-flowing rivers will want to know why they should have to reduce their waste output as much as municipalities on small, sluggish streams where the pollution problem is much greater. Again, there may be two factories, a furniture factory and a cannery, side by side, each of which dumps the same number of equivalent tons of waste into a river. Yet it may be much cheaper for the furniture factory to reduce its waste discharge by 10 per cent than for the cannery to reduce its discharge by 5 per cent. What does justice demand here? It is at least clear that the total cost of reducing waste inflow into the river from these two plants by 5 per cent would be less if the furniture plant alone were required to introduce anti-pollution measures than if both plants were required to do so.

On the other hand, if the W C B were to attempt point-by-point regulation it would have a massive administrative problem on its hands. A separate regulation would have to be drafted for each outfall, or perhaps for each polluter, each regulation taking into account the condition of the water into which the waste is discharged, and the cost to different polluters of reducing their pollution; it would also be desirable on purely economic grounds to try to minimize the total costs to all polluters of achieving a given reduction in over-all waste discharge, but to do this would require a fantastic amount of information that in practice would be very difficult and expensive to get. In any event a large bureaucracy would be required to gather all the information and undertake all the interviews that would be necessary before the thousands of individual regulations could be drafted. The administrative costs of point-by-point regulation would be enormous.

Control by regulation, then, does not seem to be a very attractive way to enforce our pollution policy. If it is of the across-the-board type it is likely to be unfair and inefficient (i.e., more costly to the polluters than some other scheme that would achieve the same reduction in pollution). If it is point-by-point regulation the whole thing seems, quite simply, impracticable. And yet the first recourse of governments when electorates demand that they "do something" about a problem is nearly always to pass laws and issue regulations about it. In Ontario, at present, most waste treatment expenditures result from laws or regulations, or from the recognition by municipalities and factories that if they don't do something about their wastes on their own initiative they will soon be forced to do so by provincial regulation. Some of the unfairness and inefficiency that we have identified as being inherent in control by regulation is reduced in practice by measures that fall between across-the-board and point-by-point schemes; regulations are applied to industries, instead of single firms, and to groups of municipalities of a certain size, or in a certain area, instead of single municipalities. For the most part, though, these compromise techniques do more to camouflage the difficulties inherent in regulatory schemes than to reduce them.

Industries and municipalities show a certain fondness for the subsidy method of controlling pollution, because they then don't have to raise *their* prices, or *their* taxes, to the consumer; it is only *provincial* taxes that are increased. This is a rather silly attitude, but it seems very hard for people to learn that *everyone* must pay to reduce pollution, and that the important question is not *who* lays out the money in the first place, but *how much* is paid to achieve what benefits. Subsidy schemes are prey to all the difficulties inherent in regulatory schemes. If an annual subsidy of so much per ton of waste withheld from the water system is given to each polluter in order to reduce his waste discharge by 5 per cent, the scheme will be unfair and also inefficient – and therefore very costly to the Treasury, because as we have seen it costs different industries and different municipalities different amounts to reduce their wastes by 5 per cent. Some will make money on the deal, others will lose. Moreover the scheme is an open invitation to blackmail, just like the scheme for paying students not to drop out of school. Once the subsidy is announced every large waste-producing industry in the country that can reduce its wastes at a cost per ton that is less than the subsidy per ton it earns by doing so will converge on the area in order to engage in the profitable industry of producing wastes and then treating them. Some of these problems could be avoided if the government subsidized everyone equally in the sense that polluters were paid, not a fixed subsidy per ton of waste reduced, but whatever subsidy was required by each polluter to reduce his waste discharge by 5 per cent. Special precautions would have to be taken to avoid fraud, but there would be no "blackmail" problem and no problem of some gaining while others lost; the scheme, however would be very inefficient, and therefore needlessly costly. If on the other hand, an attempt were made to apply the subsidy technique selectively, so that the bulk of the subsidy went to industries and municipalities that could process their wastes most cheaply, the scheme would be somewhat less inefficient in the economic sense but would run into all the problems inherent in the point-by-point regulatory scheme, namely, high costs of administration, a large bureaucracy, and long and complicated negotiations between the control agency and thousands of individual polluters.

There remains the technique of pollution charges. An across-the-board charge of so much per equivalent ton of waste discharged, levied on all municipalities and factories in a given water control region would avoid one of the major disadvantages of the other two schemes; not *all* polluters would be required to reduce their pollution by a fixed percentage, and each could decide for himself how much it would pay him to reduce his waste discharge in light of the charge that is levied. To gain this advantage, however, the W C B would have to experiment a bit with different levels of charges in order to find the one that would result in the total amount of pollution being reduced by approximately 5 per cent. However, the scheme might still be considered inefficient in that it made no allowance, at least *within* a water control region, for different intensities of existing pollution in different parts of the water system, i.e., it would make no allowance for the different waste-handling capacities of different rivers and lakes. Point-by-point charges would involve virtually all the costs, both political and economic, associated with point-by-point regulatory or subsidization schemes. It should be noted, however, that theoretical economists tend to favour the point-by-point charging scheme. Like the other point-by-point schemes it could, in principle, be completely efficient in the economic sense – though, as I hope I have made clear, I am confident that none of them could be completely efficient in practice. The point-by-point charging scheme does have some slight advantage over the others in that it requires less information to put it into effect (although it requires more trial-and-error experimentation in order to get along with the smaller amount of information); as compared with subsidization schemes it also has the "advantage" that it reduces financial problems for the government (though it increases financial problems for the polluters).

As you can see, I have tried in this section to apply a rough version of "cost-benefit" analysis to different control techniques in order to find out which is the least costly way of achieving a given benefit – or, to put it the other way around, which scheme returns the greatest benefit for a given administrative cost. And so far, the only control technique that seems to have any significant advantage over the others is the scheme of across-the-board pollution charges. So let us work on it a bit more.

A DIGRESSION ON REGION-WIDE PRICING

All of our across-the-board schemes, we have said, suffer from the disadvantage that they fail to differentiate between different locations within a given water control region. I have so far accepted the argument put forward by factory-owners, and usually accepted by economists, that of two factories discharging the same amounts of the same wastes, the one that does less damage (because it is located on a fast-flowing or very large or thinly populated body of water where the pollution problem is not serious) should pay less for pollution control than the one that does more damage (because it is located on a slow-flowing, or small, or heavily

88

populated body of water where the pollution problem *is* serious).

On second thought, I reject that argument. If it were accepted, we would be led to favour pollution control schemes that tended to even out pollution geographically and make pollution levels the same everywhere. It is true that if we could somehow equalize water pollution at, say, Belleville and Toronto, we would in some sense equalize Toronto's and Belleville's pollution damage; and it is also true that for a given average water quality (in this case the average of the quality at Belleville and at Toronto), more waste could be discharged into Lake Ontario if industry were equally divided between Toronto and Belleville than if it were unequally divided, as it now is. But I don't think people *want* pollution to be the same everywhere. As a Torontonian I sometimes go to Presqu'ile, near Belleville, to swim, and I don't *want* the swimming near Belleville to be as bad (or good) as it is near Toronto; I very much fear that if pollution levels were equalized between the two areas I couldn't swim in either. It is not because I live in Toronto that I want unequal pollution; I would feel exactly the same way if I lived in Belleville. As a matter of fact, if I valued swimming more highly than I do I would probably live in Belleville. And if a resident of Belleville valued big-city life more than he did swimming he would probably move to Toronto. The point is that we *all* benefit from variety, and that in the age of the motor car our variety doesn't have to be where we live – it can be a couple of hundred miles away. Therefore, if one of the W C B's water control regions extends from, say Toronto to Belleville, I think that it is desirable that pollution charges at Belleville should be just as high as they are at Toronto, even though Belleville has less of a pollution problem than Toronto, and thus has more "unused pollution capacity" than Toronto. That is the way I want to keep it; I want to keep Belleville less populated and less industrialized than Toronto; and I think that people in Belleville (those born there who haven't moved to Toronto, and those born in Toronto who have moved to Belleville) will agree.

The point at issue is so important that we must look at it again. The W C B, through its pollution charges, has the power to influence the location of industry. (It has a similar power to influence the location of populations; but we shall conduct the argument purely in terms of industry.) If it charges the same rates throughout an area it will have no effect on the location of industry within that area, but if there are rate differentials between cities, or different rivers, within the area, the pollution charges will have a "location effect," and the larger the differentials the larger the effect. If the Board charges less where pollution is less (Belleville) it will create a tendency for industry to move from highly polluted areas (Toronto) to less-polluted areas, and thus to spread both industry and pollution more evenly over the area. If it establishes the opposite differential, so that pollution charges are lower in highly polluted areas (Toronto), it will create a tendency to centralize both pollution and industry. Which tendency should the Board favour? Or should it adopt the neutral position of charging the same rates over an entire water control region?

We saw in chapter II that urbanization and concentration of economic activities increased pollution; on this line of thought, a policy that tends to decentralize industry seems desirable. Such a policy is sound, I think, so long as waste disposal and recreation are compatible uses of water, and up to a certain level of waste disposal the two uses *are* compatible. As was said in chapter II, if it were possible to spread the existing population and production evenly over Ontario there would probably be no pollution problem in the province.

However, the economics of concentrated production are apparently so great, and the desire to live in cities so strong, that waste disposal in urbanized areas is likely to be far greater than the amount that is compatible with at least some recreational uses of the water. At some point, then, the two uses become competitive, and as pollution grows beyond this point some recreational uses (e.g., swimming) are precluded. The effect of this development, in terms of our example, is to increase the value of the recreational uses of the water at Belleville and reduce their value at Toronto. It is therefore no longer correct to argue that industry at Belleville does less damage than it does at Toronto; this conclusion would be correct if recreation were valued at the same figure in both areas, as was reasonable so long as a full range of recreational activities was possible in both areas. But *because* of serious pollution at Toronto the recreational value of Belleville water increases and becomes greater than that of Toronto water. Thus there can be no presumption that an industry would do less damage if it located at Belleville than if it located at Toronto; it might do more.

In general, then, I do not think that region-wide pricing of pollution rights can be shown to be economically inefficient. The rather common conclusion that pollution charges should be lower in less populated and less polluted areas of a region because less damage is done in such areas seems to be based on the assumption that the fewer the people there are to be damaged in an area the less the damage that will be done. This argument overlooks the elementary point that people are mobile and do not stay at home every hour of the year. In fact, most of the damages done by pollution in lightly populated sections of Ontario are likely to be suffered, not by the residents of those areas, but by city people, who place higher and higher values on unpolluted areas within easy driving distances of their homes as their urban environments become more and more polluted.

As you may have realized, we have now come, by a rather roundabout route, to the same conclusion we reached in chapter v, namely, that a "separate facilities" solution to conflicting interests – such as pollution versus recreation – is often desirable. Residents of East Toronto and West Toronto cannot separate the water they use, and they must come to a *common* decision about what the quality of water at Toronto is going to be (or accept the decision of some higher-level government). But there is no necessity for the quality of water at Belleville to be the same as the quality of water along the Toronto beaches. There is every reason, it seems to me, to try to keep it different. Let us hope that provincial control of pollution will not result in province-wide pollution! We have argued our point in terms of different areas within one water control region. But exactly the same argument applies to the question of pollution charges as between the regions themselves.

We cannot presume to advise the real authorities, but we can

91

issue directives to our imaginary W C B. I suggest that we tell the Board that it must never adopt a system of pollution charges that will tend to decentralize industry and spread pollution around; pollution in one region must never be reduced by increasing pollution in another. The Board may adopt a locationally neutral pricing policy – which implies the same price per ton of waste discharged throughout the whole province; or it may charge different rates in different water control regions, so long as the lower rates are in the *more* polluted areas and the higher rates in the *less* polluted areas. The locational effect of the differential charges will then be to concentrate industry and pollution even further in the areas where they are now concentrated, and to create a greater contrast in pollution levels between different areas than now exists. These are strong directives, and we may wish to modify them at some later date in the light of experience. At the moment, however, we opt for a pricing system that favours differential pollution between areas and regions rather than one that tends to equalize pollution everywhere.

THE BEST WAY

Once we drop the charge of economic inefficiency against region-wide pricing, the across-the-board pollution-charges technique of implementing our pollution policy looks even better. We have already seen that it is efficient as between different firms and municipalities because each polluter decides for himself by how much, if at all, he should reduce his wastes. The burden of pollution control is thus shared in exactly the right way, without the W C B's having to agonize over the question of how to find a just and reasonable sharing of the cost of the scheme. Since every polluter adjusts to the charges in whatever way minimizes *his* cost, the social cost of achieving the target amount of waste discharge – which is the sum of the costs borne by all polluters (and, of course, by the consumers of their products, in the case of industries) – will also be automatically minimized. There can be no doubt that this scheme is by far the most efficient, i.e., the least costly, way for the W C B to implement its policy. It has the great additional advantage, as compared with any of the point-by-point schemes,

that is administrative costs are very low. The main thing the Board
has to do to put the scheme into effect is to declare the fee that
everyone must pay for the right to discharge one equivalent ton
of waste into natural waters anywhere in a water control region.
(There are some other things it will have to do also, but we shall
discuss the administrative problem in more detail presently.)

There remain, however, two awkwardnesses in the scheme:
the trial-and-error procedure that is necessary before the Board
can hit on the "right" level for the pollution charge; and the
"guesstimate" it must make about how much existing polluters
should reduce their wastes in order to allow new-comers (people
or factories) to settle in the region without increasing the *total*
amount of waste discharged into the water system. Can these
shortcomings of the scheme be remedied? I think they can.

MARKETS IN POLLUTION RIGHTS

Let us try to set up a "market" in "pollution rights." The Board
starts the process by creating a certain number of Pollution
Rights, each Right giving whoever buys it the right to discharge
one equivalent ton of wastes into natural waters during the current
year. Suppose that the current level of pollution is roughly satisfactory. On this assumption, if half a million tons of wastes are
currently being dumped into the water system, the Board would
issue half a million Rights. All waste dischargers would then be
required to buy whatever number of Rights they need; if a factory
dumps 1000 tons of waste per year it will have to buy 1000 Rights.
To put the market into operation, let us say that the Board decides
to withold 5 per cent of the Rights in order to allow for the growth
of production and population during the first year, and therefore
offers 475,000 Rights for sale. Since the demand is for 500,000,
the Rights will immediately command some positive price – say,
10 cents each.

Even at 10 cents per Right some firms will find it profitable to
treat their raw wastes before they discharge them, or to dispose
of them in some way other than discharging them into the water.
They will thereby reduce the number of Rights they are compelled
to buy, and, when the price has risen enough to reduce the demand

93

by 25,000 Rights, the market will be in equilibrium. As time goes on, we would expect the growth of population and industry to result in an increase in the demand for Rights, and since the number of Rights issued by the Board cannot be increased the price of the Rights will move upward. As it does so, the incentive for waste dischargers to treat, or reduce, their wastes, so that they reduce the number of Rights they must buy, increases.

Once the market is in full operation, individual holders will buy and sell Rights on their own initiative, but always through the one "broker," the w c b. Firms that go out of business during the year, or that experience a slump in production, or that bring new waste disposal practices into operation, will have Rights to sell; new firms, or those that find that their production is exceeding their expectations, will appear as buyers in the market. Similarly for municipalities; those that build new or better sewage treatment plants will be sellers of Rights, while those that experience growth in their populations and do nothing to reduce their wastes will be buyers. All of these buyers and sellers, through their bids and offers, will establish the price of the Rights. The price will no doubt display minor fluctuations from time to time, like other prices; but it will probably show an upward trend over time. That makes sense; if economic growth (which causes pollution) is to continue, and yet pollution is to be checked, the cost of disposing of wastes must rise – and this increasing cost is registered in the rising price of Pollution Rights.

Like all organized markets, our Rights market must be conducted according to certain rules. As we have seen, a person who wants to sell some of his Rights (because he no longer needs them) can sell them to somebody who has a bid in for Rights; but if it happens that there is no bid at the moment, or if the bid is much below (say, more than 10 per cent below) what the seller paid for his Rights, the w c b should stand ready to buy the unexpired portion of them, at, say, 90 per cent of their purchase price. In this way the w c b acts like a "specialist" on organized stock exchanges; as a buyer of last resort, the Board prevents any sudden fall in price that might occur, for example, if several large industries more or less simultaneously introduced waste control measures and thus temporarily flooded the market by wanting to sell large numbers of Rights. In the same way, when at the beginning of the next year old Rights are extinguished and new ones put up for sale, the Board must "rig the market" so as not to let the price fall very much; otherwise municipalities and factories that had introduced treatment procedures when the price of Rights was high might find that their investments were unprofitable at the new, lower price.

In general, though, the typical problem, if any, will be that of sharp price *rises*. The Board should try to avoid these by keeping a certain reserve of "issued but unsold" Rights on hand, and selling some of this reserve supply if demand shows a sharp rise that is expected to be only temporary. The Board should be very firm, however, in refusing to increase its authorized issue *no matter what happens*. If municipalities or factories ever got the idea that by complaining loudly enough about their inability to buy Rights ("at any price" as they will likely put it) they could get the Board to increase its issue, even "temporarily," the Board's basic pollution-control policy would be shattered; moreover, by issuing "excess" Rights the Board would be breaking faith with the owners of Rights by "diluting their equity," and preventing them from selling their Rights, if they wished to do so, at as large a profit as they otherwise could have gained.

So far we have argued as if all Rights should be for one year only. It would, in fact, be desirable to issue Rights of different terms – up to five-year Rights if the Board has announced that its present policy (the number of Rights it will issue) is to be in force for five years. Long-term Rights would no doubt command a premium over short terms – a three-year Right would cost more than three times as much as a one-year Right – since the firm (or municipality) that bought it would be secure in its discharge rights for three years rather than one year. Different dischargers will probably want to buy "security of discharge" for different lengths of time into the future, and there is no reason why they shouldn't do so.

Anyone, of course, should be allowed to buy Pollution Rights, even if they do not use them. Conservation groups might well want to buy up some rights merely in order to prevent their being used. In this way at least part of the guerilla warfare between conservationists and polluters could be transferred into a civilized "war with dollars"; both groups would, I think, learn something in the process. Pure speculators should also be able to buy the Rights in the expectation of being able to sell them later at a higher price—and also to sell them short if they think the price will go down in the future. Speculation is a risky business for the speculators, but it does help to "make a good market," and if enough speculators can be found to play the Rights market they will help to even out temporary price fluctuations and thus help the Board stabilize the market. As experience is gained in running the market, other rules might become desirable, just as it is often found desirable to change the rules for trading on a stock exchange. But we have perhaps established the main rules needed to get the market into operation.

How would the Government use the money that the W C B took in by selling pollution rights? In any way the government sees fit. Let us say it goes to consolidated revenues. There is no problem in disposing of money!

Once in operation, the Pollution Rights market will, by establishing a price for Rights, relieve the Board of any necessity to *set* the proper price by trial-and-error methods. The market will also automatically solve the problem of new-comers. As population and factories grow, the price of Rights will automatically rise, and existing polluters will find it profitable to reduce their own wastes in order to sell some of their existing Rights at a profit or in order to avoid buying so many of them next year; reduced demand by existing holders will release a supply for new buyers.

At the end of the initial five-year period, the Board may wish to revise its policy about the amount of waste it will allow to be discharged into the waters of its various water control regions. Revision of policy, if any, will simply mean the authorization of different (higher or lower) maxima for the number of Rights that can be issued in each Region. At this point, all the pressures of public opinion, political considerations, and interest-group propaganda, will converge on the Board; but the Board, so long as it enjoys the confidence of the government and so long as the government enjoys the confidence of the electorate, must make its own decision. And once it decides what to do, it can continue to use the pollution markets to do it.

ADMINISTRATIVE COSTS

Any pollution-control scheme will require a certain amount of policing. In the scheme we have recommended it would be necessary to ensure that waste dischargers did not cheat by buying too few Rights (or discharging too much waste). The W C B should certainly maintain a system for measuring the amounts of different wastes in the water and in the effluents from outfalls, not only to check upon possible fraud but also to find out how well its policy is working. (We must remember that pollutants in surface run-off have not yet been brought into the orbit of Board policy.) It should also be possible to calculate approximately how many equivalent tons of waste are likely to be generated by a given population or by a factory producing a given output of a certain product, and if certain municipalities or industries are buying a much smaller number of Rights than their estimated tonnage of wastes, the Board would probably want to look into the discrepancy. But the Pollution Rights scheme, it seems clear, would require far less policing than any of the others we have discussed; the setting of regulatory standards, for example, would probably involve almost continuous monitoring of effluents to ensure that the standards were not being breached.

Moreover, it is obvious that the across-the-board pollution-charges scheme would involve only a small fraction of the administrative personnel and expense that would be required to administer any point-by-point scheme; there would be no need to employ an army of people to draw up hundreds or thousands of standards, or to arrange individual subsidies, or to set damage fees on a plant-by-plant, municipality-by-municipality basis. The Pollution Rights market will also, as we have seen, simplify administrative problems by removing the necessity of anyone's deciding what pollution charges should be in order to achieve the Board's goal and to provide for economic growth without increasing the total amount of pollution. The administrative simplicity of the scheme is certainly one of its main attractions.

97

RUN-OFF POLLUTION

Wastes that enter water and air systems not at identifiable outfalls or "emission points," but at hundreds of thousands, or millions, of points, pose special problems. We are not yet sure how important run-off wastes are in causing water pollution, but there is ample evidence that private automobiles are a major source of air pollution in cities.

I shall not comment extensively on these problems. Pollution Rights markets don't seem practicable. Across-the-board regulatory schemes (which involve a change in property rights) seem attractive where they can be easily enforced, as is the case in proposals to require automobile producers to equip their products with emission-control devices. (It is quite another matter, however, to police the effectiveness of such devices that have been in use for several years.) Farmers located along lakes or rivers might be required to take measures to prevent excessive run-off from their land. Excise taxes – a form of across-the-board system of charges – might be used to increase the cost of, and therefore to encourage a more efficient use of, fertilizers and insecticides; a licence fee for automobiles that was more steeply graduated with respect to horsepower than is now the case might very well do more to reduce air pollution than regulations requiring emission-control devices. Subsidies may be justifiable in some cases, as where a farmer has a run-off problem that is particularly difficult to correct.

In giving such brief attention to "diffuse" pollution, I do not imply in any way that it is unimportant. It is often very important, and in my opinion ought to be attacked vigorously. A number of problems are involved, and no doubt it will be desirable to utilize a variety of techniques for controlling them.

SUMMARY

A pollution control policy, in the commonly accepted sense of the phrase, must control pollution. Someone, somehow, has to come up with some sort of limitation on the dumping of wastes into the water and the air. Or, to put it another way, whoever "owns" the air and the water must establish rules about their use or what we called in chapter v "property rights" in their use. There are an infinite number of such policies or rules about property rights. In this chapter we have arbitrarily chosen one water-pollution control policy for discussion purposes. The policy consists simply of dividing up a large piece of geography, Ontario, into several smaller pieces of geography, which we called water control regions (I haven't said how many water control regions there should be because I don't *know* how many there should be) and having an authority, a water control board, decide the maximum number of equivalent tons of waste that can be dumped in the waters of each region in each year.

The main purpose of this chapter has been to show that economic analysis, which is all but useless in helping us to decide on a policy, is all but indispensable in helping us to decide on the best way of implementing a policy once it has been chosen. The criterion is simply that the best way of implementing a policy is the least costly way, counting *all* costs: costs to individuals as provincial tax-payers (including the very important cost of administering the scheme); costs to individuals as municipal tax-payers; and costs to individuals as consumers of goods and services.

On this criterion, the system of charging a uniform amount for the right to discharge wastes into the natural water system (so much per equivalent ton of waste per year to every polluter at any location in the region) was found to be far superior to any of the other half-dozen schemes considered. It was efficient in the sense that the total direct cost of implementing the policy was distributed amongst waste-dischargers in the fairest and least costly possible way; each polluter decided for himself how he could minimize his costs of waste disposal – to what extent it would be profitable for him to reduce his wastes and to what extent it would be profitable to continue to discharge them into the natural water system even after paying the discharge fees or pollution charges of doing so. It was the most efficient scheme also in the sense that its administrative costs appeared to be lower than those of any other scheme. Moreover, administrative costs could be still further reduced by establishing full property rights to use water for waste disposal up to a limit set by the authorities, that is to say by setting up Pollution Rights markets. Such markets would automatically set the correct level of the pollution charge (instead of its having to be set by some committee, after long and learned discussion) and would also automatically, and continuously, adjust the level of the charge to take account of economic growth. A simple market that can be operated by three or four people and a small staff of stenographers to register purchases and sales is very much cheaper, and just as efficient, as a large bureaucracy replete with computers to give answers to complicated pricing problems. If it is feasible to establish a market to implement a policy, no policy-maker can afford to do without one. Unless I am very much mistaken, markets *can* be used to implement any anti-pollution policy that you or I can dream up.

The only real question is how much you and I are willing to pay in order to reduce pollution.

Summary

Pollution is popularly conceived as a question of polluters *versus* pollutees, of "them" against "us." In some particular instances this is certainly a correct view of the problem. Thus if a pulp and paper mill locates near a vacation area and disposes of its wastes in a way that seriously reduces the local amenities, tourist operators in the area will certainly be damaged by the mill-owners' actions. In such cases the two parties are economically quite separate (the tourist operators buy nothing from the mill, which buys nothing from the tourist operators), and their interests are conflicting. If the mill has a legal right to use the water for waste disposal purposes, the tourist operators will be unable to obtain compensation for the damages they suffer; all they can do is to campaign for a change in the definition of property rights to the water in question.

The main purpose of this volume, however, has been to put pollution into the perspective of a society-wide problem. In the vast majority of pollution problems, it seems to me, we pollute each other, so that it makes little sense to classify people into polluters on the one hand and pollutees on the other. In a city, everyone who travels by car or bus or uses an electrically driven public transportation system that receives its power from fuel-burning generating stations, contributes to the pollution of the air he and others breathe. If factories in a metropolitan area pollute regional waters, most people in the area will consume the factories' products or work for the firms concerned, or own stock in them. And if some of the manufactured products are exported from the area, so that the residents bear the pollution associated with consumption by others, they also import goods from others who bear the pollution associated with their export production. If others' sewage pollutes us, our sewage probably pollutes others. In pollution matters, in brief, we usually do about as much damage as we suffer. That is why pollution is a *social* problem and why we must decide *collectively* what to do about it.

The emotional heat generated by popular discussions of pollution derives, I think, from two suspicions – that pollution is dangerous to health, and that it may upset the balance of Nature to such an extent that the planet could become uninhabitable – and one observation, namely, that the quality of our natural environment is deteriorating from an aesthetic point of view. I can offer only a layman's commentary on these matters, since I am neither a doctor nor a biologist nor a philosopher; but I make no apology for the following remarks because it is we laymen who, in the final analysis, must decide what, if anything, is going to be done about pollution.

The health danger resulting from water pollution is minimal, in my opinion. The existing technology of water treatment seems adequate to provide good-quality drinking water from the most polluted of natural waters, and the prohibition of swimming in polluted areas reduces the public health hazard of water-borne disease to small proportions. A majority of the earth's population probably lives in the midst of severe water pollution, and yet manages not only to survive but to increase. The canals of Venice and Amsterdam are highly polluted, and yet cause no serious problem of public health; and Mexicans seem to have adjusted to water supplies that are quite likely to upset Americans and Canadians.

Air pollution is quite another matter. There can be little doubt that some types of pollutants, in certain concentrations, are detrimental to health. There is at present very little firm knowledge about the dangers involved, but research effort in the field is being greatly increased and we should know more about the health risks of air pollution in a few years. Even now, however, it seems quite clear that smoking, especially cigarette smoking, is at least as damaging to health as present levels of air pollution in large cities. The human respiratory system seems able to cope satisfactorily with a great deal of air pollution, and recent studies suggest that it is mainly when the system is burdened with both pollution *and* smoking that a high risk situation is created. To that extent, one's respiratory health, even in large cities, seems to be largely under one's own control. On the other hand, everyone would presumably be in a lower risk category if we were to insist on lower levels of air pollution. Our lack of precise knowledge about the health hazards of polluted air should surely, in itself, lead us to be doubly cautious about levels of air pollution. It must be remembered that the air in most North American cities has probably been seriously polluted only during the past thirty years, so that present statistics relate to a thirty-year exposure to pollution; what effect sixty years of exposure to pollution will have on human lungs is, I suppose, anyone's guess.

In recent years, scientists have pointed to the danger that the rapidly increasing combustion of fuels may result in a significant depletion of the world's oxygen supply, and to the possibility that the exhausts of high-flying supersonic aircraft may lead to an increasing carbon-dioxide cloud cover over the world, and a consequent rise in temperature sufficient to melt the polar ice caps and flood coastal populations. Among other horrors that we live with are the unknown, but almost certainly not beneficent, effects of increasing concentrations of DDT and other chlorinated hydro-carbons in human and animal tissues, and the possible escape of radioactive wastes in quantities that could kill millions of us and radically alter the chemical, physical, and biological nature of the world. Such eventualities are literally unthinkable. They smack of science fiction, but no one who is aware of how rapidly science fiction has been translated into reality during the recent past can completely dismiss these dire possibilities.

The only wholly tenable argument against pollution at the present time is the aesthetic one. It is fool-proof because if you and I say that polluted air and water and soil are aesthetic insults we cannot be proved wrong by logical argument. (We, on the other hand, cannot prove wrong those who think that pollution is a fine thing or at least a necessary evil; but we *can* question their sense of values and their taste.) And surely our position is reasonable. Take a plane trip on a clear day; notice the brown-grey palls over our cities, and the stains on the waters in their vicinities. No affluent society should permit such outrages; neither, perhaps, should poor societies, but the poor have less choice. I would gladly blame pollution on the politicians, and accuse them of moral cowardice in pandering to public demands for cheap goods and low taxes; but I am given pause by the reflection that you and I are the public, and that we may very well be the authors of our own misfortunes. On second thought, though, I do blame the politicians: after all, they set themselves up as *leaders*.

We all know that pollution can't be wished away, and that waste disposal costs are as inevitable as death and taxes. We must bear these costs in one of three forms: treating wastes to reduce their noxious effects *before* we release them into the environment; avoiding their noxious effects *after* they have been released into the environment; or simply *suffering* their noxious effects after they have been released into the environment. The economist, from his Olympian heights, says that what we should do, obviously, is to allocate costs between pollution prevention, damage avoidance, and welfare damage in such a way as to minimize the social costs of waste disposal.

But the economist cannot tell us how to allocate our costs of waste disposal, because he cannot measure welfare damages. And if he cannot measure this term in the equation, he cannot use the differential calculus to tell us what the values of the other three terms *should* be. All he has to offer, therefore, so far as an anti-pollution *policy* is concerned, is a counsel of perfection. In the end it is you and I who are going to elect the politicians who are going to decide how much pollution we are going to have, what sort of pollution it is going to be, and where we are going to have it. If you are concerned about present levels of pollution, insist that your governments (which own the air and the water, and moreover determine all property rights) start reducing the amount of pollution. And don't let them fob you off with a royal commission or a task force on the grounds that economists (or any other species of expert) will be able to tell them exactly the right amount to spend on pollution prevention. No one knows the answer to that question. The politicians must decide what the public wants and stake their political lives on their decision; they are in a much better position to assess the benefits and costs of their action (or inaction) than any body of experts.

Politicians are, presumably, experts in the art of social policy-making; but they cannot be considered experts in policy implementation. In this essay I have tried to draw a clear distinction between making a policy and implementing it, yet I realize that the interrelations between a policy and at least the main strategy of its implementation are so close that a good decision about policy goals may easily be rendered ineffective by an inferior strategy of implementation. The government, and especially the cabinet, must therefore be responsible for strategic decisions about the implementation of their policies; but since they are not experts in this field they must rely on advice from others if they are to come to a wise decision on the matter.

I have suggested that for many social policies, and for a pollution policy in particular, there are three main strategies of implementation: regulation; subsidization; and some system of charging for user rights. My own thinking on these matters leads me strongly to favour the charging system wherever practicable. I must stress, however, that my arguments and conclusions on this subject are tentative; I feel very strongly that social scientists have not devoted nearly enough effort to the discussion of implementation problems, and that it will be some time before anyone can write on this question with much confidence. The question is complex because it is an amalgam of legal, economic, and administrative matters, and lawyers, economists, and students of public administration have not yet co-ordinated their attacks on it.

It was argued in chapter v that water and air can no longer be

105

reasonably treated as a common property resource open to anyone to use as they see fit, because such a policy will lead to continuous depreciation of the quality of these valuable assets. Systems of limiting access to the resource that depend on regulation or subsidization are similar except that the money cost of meeting regulations is borne in the first place by the discharger (and then reflected in higher prices of goods and higher municipal taxes), while under a scheme of provincial subsidies part of the money cost is reflected in higher provincial taxes. Both systems – regulation and subsidization – create a sort of property right that we may call "fixed tenure" or "certified tenure." That is to say, once the discharger has obeyed the regulation (or introduced a subsidized treatment system) he has earned some sort of right to discharge the rest of his waste into the water system; the authorities give him, so to speak, "a certificate of satisfactory performance." This right, however, is not a full property right because it cannot be transferred; it inheres in the particular discharger who has earned it or been granted it. It is true that if a private discharger comes to sell his factory, the value of the right may be reflected in the selling price. Our present society is full of such implicit prices – the value of a taxi license, of a milk-collecting route, of the right to grow tobacco, of tariff protection – all of them reflecting the value to particular parties of some governmental regulation. This situation is unsatisfactory in at least two respects: the right is not always clearly defined, and it is not always clear that the vendor can give the purchaser clear title to it; moreover, the price is a capitalized value, transactions are infrequent, large sums of money are frequently involved, and the public, when it learns of them, is not convinced that it is fair that a private party should be able to capture the capitalized value of a publicly conferred right. (Of two farms suitable for growing tobacco, the one with a "quota" will sell for two or three times the price of the other.)

To an economist, however, the chief disadvantage of regulatory schemes and point-by-point subsidization schemes is that they cannot possibly be efficient; to draw up a list of regulations or subsidies that would reduce pollution by, say, 10 per cent *and do*

106

so in such a way as to minimize the cost of the operation, is humanly impossible. It amounts to asking a government agency to solve a set of several thousands of simultaneous equations when it doesn't even have all the information necessary to write down the equations in the first place. An across-the-board subsidy scheme of a flat rate per ton of waste discharge reduced would avoid this difficulty, but it would not avoid two other disadvantages of regulation and subsidization systems. If water (or air) pollution is to be prevented from exceeding a given figure, then restrictions on use will have to become more severe as population and production grow; thus as economic growth proceeds, regulations or subsidies will have to be continuously changed. Moreover, it seems obvious that both the setting up of such schemes and their continual revision to take account of economic growth will require heavy administrative costs and large bureaucracies.

The solution recommended for consideration in this essay is designed to avoid the disadvantages of the schemes we have been discussing. It is suggested that transferable property rights be established for the disposal of wastes. The government can choose any level of pollution it wishes by setting the number of regional Pollution Rights it issues – the number to be subject to change at five-year or ten-year intervals. Because transferable (or full) property rights always command an explicit price, the establishment of such Rights makes it easy to establish a market in them. In turn, the buying and selling of the Rights in an open market and the consequent establishment of an explicit price for the right to discharge a ton of wastes into a water (or air) system results in a theoretically efficient allocation of "anti-pollution effort" as between different dischargers. In other words, the market automatically ensures that the required reduction in waste discharge will be achieved at the smallest possible total cost to society. Moreover, the rise in the price of the Pollution Rights over time will automatically solve the problem of economic growth; as the price rises, it will be economic for existing dischargers to reduce their wastes, and thereby make room for newcomers. And, finally, it seems obvious that the Pollution Rights

107

market will require very little administrative expense by comparison with alternative schemes.

The major deficiency in this solution is that it would be impracticable to use it to reduce "multiple source" pollution – general run-off of insecticides, herbicides, and fertilizers from farms, or emissions from automobiles and domestic heating plants in the case of air pollution. Here a variety of regulations (as in automobile exhaust devices), subsidies (to build earthworks to reduce run-off from farm land, for example), and excise taxation (to reduce the careless over-use of insecticides, herbicides, and detergents, for example) seems appropriate.

Economic growth, population growth, and urbanization are the great generators of pollution. In the final analysis, the pollution problem could be brought under control by halting the growth of these statistics. One economist has recently expressed his personal opinion that something of the sort is required if we are to keep our pollution and congestion problems at bay. I can understand his concern and his passion. And I agree with him that we have been ridiculously "oversold" on the virtues of growth in the gross national product; such a goal, in my opinion, is absurd and not worthy of free men. But I think we still have time to make our environment livable without going to extremes. The taming of pollution problems is not really very difficult. It will cost a lot of money, but we are very rich, and we really don't want to "live off our capital" – especially our very important assets of air and water. All it requires is a resolve on our part to trade off a modest part of our goods-and-services standard of living for an improved environment – and communication of this resolve to our politicians in no uncertain terms.

Some further reading

Information on pollution problems in Canada is contained in mimeographed collections of papers presented at three recent conferences. *Pollution and Our Environment* is a large volume of papers given at a conference convened by the Canadian Council of Resource Ministers and held at Montreal from October 31 to November 4, 1966. The papers from the *Ontario Pollution Control Conference*, held at Toronto, December 4–6, 1967, have been published by the Ontario government. They present a useful survey of pollution problems and controls in Ontario. The long quotation in my chapter v comes from an article on "Fish and Wildlife Values in Pollution" by C. H. D. Clarke published in the Ontario volume (pp. 149–151). *Proceedings of Great Lakes Water Resources Conference* (June 24–26, 1968, Toronto) is available from the Engineering Institute of Canada; the papers under the general headings of "Water Use Conflicts" and "Water Quality in the Lakes" are relevant to pollution problems.

Scientific information on the Great Lakes is published by the Great Lakes Institute of the University of Toronto and in annual volumes of *Proceedings* of the *Conference on Great Lakes Research* sponsored originally by the Great Lakes Research Division of the University of Michigan and the Great Lakes Institute of the University of Toronto, and since 1967 by the International Association for Great Lakes Research. In the present book I have drawn information from P. F. Hamblin and G. K. Rodgers, *The Currents in the Toronto Region of Lake Ontario*, Publication PR29 of the Great Lakes Institute (mimeographed, 1967);

and from G. K. Rodgers and D. V. Anderson, "The Thermal Structure of Lake Ontario" in *Proceedings of the Sixth Conference, Great Lakes Research* (University of Michigan, Great Lakes Research Division, Publication 10, 1963, pp. 59–69). Several articles, among them those by D. Misener and G. K. Rodgers, in Claude E. Dolman, ed., *Water Resources of Canada* (Royal Society of Canada, Studia Varia no. 11, Toronto, 1967) make interesting and informative reading.

The orthodox economic approach to water pollution problems is very well presented in Allen V. Kneese and Blair T. Bower, *Managing Water Quality* (Baltimore, 1968); this volume also contains an interesting account of the organizations for pollution control that have developed over a long period in the Ruhr valley of West Germany. Discussions of the advantages and shortcomings of "benefit-cost" analysis may be found in two articles in the *Economic Journal*: "Economics in Unwonted Places" by B. R. Williams in the March, 1965, number, pp. 20–30; and "Cost-Benefit Analysis: A Survey" by A. R. Prest and R. Turvey in the December, 1965, number, pp. 683–735. An amusing book that shows how economists think about problems in general is Robert A. Mundell's *Man and Economics* (Chicago, 1968).

An unorthodox approach to pollution problems, and several other contemporary social problems, may be found in a recent book by an economist with a high reputation in orthodox economics – *The Costs of Economic Growth*, by E. J. Mishan (New York, 1967). I have drawn especially on chapter 8 of this book, where Professor Mishan discusses "separate facilities" solutions to social problems. Professor Mishan also discusses property rights, and indeed recommends the establishment of a novel form of such rights, "amenity rights" to clean air and quiet surroundings. (In my view it is more practicable, at least in Canada, to establish rights to make air and water dirty rather than to keep them clean; but both systems amount to much the same thing, since the fewer the rights that are granted to dirty our environment the cleaner our environment will be!) Professor Mishan is the economist referred to in the last paragraph of my book.

There are three basic articles on property rights that may yet serve to promote a creative fusion of economics, law, and political science (especially that branch of political science known as public administration). The first is by Professor H. S. Gordon, "The Economics of a Common-Property

Resource: The Fishery," in the *Journal of Political Economy*, April, 1954, pp. 124–42; the second is by Professor Ronald Coase, "The Problem of Social Cost" in the *Journal of Law and Economics*, October, 1960, pp. 1–44; and the third is "The New Property" by Professor Charles A. Reich in the *Yale Law Journal*, April, 1964, pp. 733–87. The Gordon article shows the wastefulness of common-property arrangements; the Coase article is eloquent on the relationship between law and economics; and the Reich article illustrates profusely the great variety of imperfectly transferable property rights that result from governmental regulations and court decisions. My chapters 5, 6, and 7 owe much to these three articles. Taken together, these articles are, I believe, pregnant with ideas that help us understand many vexing problems in "social economics" – among them, pollution problems.